GEOINFORMATION
Remote Sensing, Photogrammetry, and Geographic Information Systems

SECOND EDITION

**Gottfried
Konecny**

CRC Press
Taylor & Francis Group
Boca Raton   London   New York

CRC Press is an imprint of the
Taylor & Francis Group, an **informa** business

COVER IMAGES: The Himalayas from Nepal to Mount Everest to Tibet, and East Germany in freshly fallen snow. Spacelab 1 Images. The first European manned mission launched in 1983 by NASA and ESA that carried a Zeiss Metric Camera delivering high-resolution space images with 10 m ground resolution in stereo. The mission was suggested by the author Gottfried Konecny, who served as Project Scientist.

CRC Press
Taylor & Francis Group
6000 Broken Sound Parkway NW, Suite 300
Boca Raton, FL 33487-2742

© 2014 by Taylor & Francis Group, LLC
CRC Press is an imprint of Taylor & Francis Group, an Informa business

No claim to original U.S. Government works

Printed on acid-free paper
Version Date: 20140127

Printed and bound in India by Replika Press Pvt. Ltd.

International Standard Book Number-13: 978-1-4200-6856-6 (Hardback)

Library of Congress Cataloging-in-Publication Data

Konecny, Gottfried.
  Geoinformation : remote sensing, photogrammetry and geographic information systems / Gottfried Konecny. -- Second edition.
      pages cm
  Includes bibliographical references and index.
  ISBN 978-1-4200-6856-6 (hardback)

  1. Geographic information systems. 2. Remote sensing. 3. Photogrammetry. I. Title.
  G70.212.K65 2014   910.285--dc23

Visit the Taylor & Francis Web site at
http://www.taylorandfrancis.com

and the CRC Press Web site at
http://www.crcpress.com

GEOINFORMATION

Remote Sensing, Photogrammetry, and Geographic Information Systems

SECOND EDITION

Contents

Preface

In the 1990s, surveying and mapping underwent a transition from discipline-oriented technologies, such as geodesy, surveying, photogrammetry, and cartography, to the methodology-oriented integrated discipline of geoinformatics. This is based on the Global Navigation Satellite System (GNSS), or GPS, positioning, remote sensing, digital photography for data acquisition, and a geographic information system (GIS) for data manipulation and data output. This book attempts to present the required basic background for remote sensing, digital photogrammetry, and GIS in the new geoinformatics concept in which the different methodologies must be combined.

For remote sensing, the basic fundamentals are the properties of electromagnetic radiation and their interaction with matter. This radiation is received by sensors and platforms in an analogue or digital form, and is subjected to image processing. In photogrammetry, the stereo concept is used for the location of information in 3D. With the advent of high-resolution satellite systems in stereo, the theory of analytical photogrammetry restituting 2D image information into 3D is of increasing importance, merging the remote sensing approach with that of photogrammetry. The result of the restitution is a direct input into geographic information systems in vector or raster form. The fundamentals of these are described in detail, with an emphasis on global, regional, and local applications. In the context of data integration, a short introduction to the GPS satellite positioning system is provided.

This book will appeal to a wide range of readers from advanced undergraduates to all professionals in the growing field of geoinformation.

Acknowledgments

I would like to acknowledge the support I received from a number of individuals. My successor in office at the University of Hannover (Germany), Dr. Christian Heipke, supported me within the institute's infrastructure. To Irma Britton of the publisher, Taylor & Francis, I would like to express my special thanks for her encouragement and her patience to have the second edition of this book completed. Professor Reinartz of DLR, Jack Dangermond of ESRI, and Victor Adrov of Racurs, made it possible to add many new illustrations to the book.

On a personal level, I would like to thank my wife, Lieselotte, who has also been very patient with me and who has fully stood behind me in my activities.

Author

Gottfried Konecny is emeritus professor at the Leibniz University of Hannover in Germany and former president of the International Society of Photogrammetry and Remote Sensing.

GEOSPATIAL WORLD LEADERSHIP AWARD

On the Occasion of Geospatial World Forum 2013

Geospatial World Magazine is proud and privileged to acknowledge

Professor. Gottfried Konecny

for Lifetime Achievement

An academic, a passionate teacher, a researcher and a visionary is Dr. Gottfried Konecny, the Emeritus Professor and former Director of the Institute of Photogrammetry, Leibniz University, Hannover, Germany.

Prof. Konecny has been associated with Photogrammetry and Geoinformation since 1945 when he began his career in the Survey Office at Troppau in Czechoslovakia. He setup the Department of Surveying Engineering at University of New Brunswick, Canada offering the first English speaking undergraduate and graduate degree program in Canada for the subjects of Surveying, Geodesy, Photogrammetry and Cartography where he continues to be the Adjunct Professor since 1971. He then, took over as the Director of Institute for Photogrammetry and Engineering Surveys, University of Hannover, FR. Germany, responsible for teaching, research and consulting activities since 1971 Prof. Konecny has received some of the highest honours from the world's societies in cartography, surveying, mapping, photogrammetry and remote sensing. He has been conferred Honorary Doctorates from three internationally acclaimed Universities. He has been the recipient of numerous scholarships like the Fulbright and the NSF Fellowship from USA, the USSR Academy of Sciences Fellowship and the Commonwealth of Australia - Vice Chancellors Fellowship. He was, for his dedicated services, conferred the Order of Merit, First Class, by the Federal Republic of Germany in 1990. For the last forty years, Prof. Konecny has effectively used his vision, thought and energies in bringing synergy between the academic community specialising in the geoinformation sciences, and the geospatial industry, by being a leading example conducting meaningful research, development and capacity building. A number of developing countries in Asia owe the initiation of academic centers of excellence in geoinformation to Prof. Konecny.

C I T A T I O N

15ᵀᴴ MAY 2013
ROTTERDAM
THE NETHERLANDS

DR. M P NARAYANAN
CHAIRMAN
GEOSPATIAL MEDIA AND COMMUNICATIONS

SANJAY KUMAR
CEO
GEOSPATIAL MEDIA AND COMMUNICATIONS

A copy of the citation of the Lifetime Achievement Award by the Global Geospatial Forum in Rotterdam, 2013.

List of Figures

Chapter 1

Introduction

Surveying and mapping in the 1990s underwent a transition from discipline-oriented technologies, such as geodesy, surveying, photogrammetry, and cartography to a methodology-oriented integrated discipline of geoinformation based on Global Navigation Satellite System (GNSS), or GPS, positioning, remote sensing, digital photography for data acquisition, and a geographic information system (GIS) for data manipulation and data output. This book attempts to present the required basic background for remote sensing, digital photogrammetry, and geographic information systems in the new geoinformation concept, in which the different methodologies must be combined depending on efficiency and cost to provide spatial information required for sustainable development. In some countries this concept is referred to as "geomatics."

For remote sensing the basic fundamentals are the properties of electromagnetic radiation and their interaction with matter. This radiation is received by sensors on platforms in analogue or digital form to result in images, which are subject to image processing. In photogrammetry the stereo concept is used for the location of the information in three dimensions. With the advent of high-resolution satellite systems in stereo, the theory of analytical photogrammetry, restituting two-dimensional image information into three dimensions, is of increasing importance, merging the remote sensing approach with that of photogrammetry.

The result of the restitution is a direct input into geographic information systems in vector or in raster form. The application of these is possible at the global, regional, and local levels.

Data integration is made possible by geocoding, in which the GPS satellite positioning system plays an increasing role. Cost considerations allow a judgment on which of the alternate technologies can lead to an efficient provision of the required data.

SURVEYING AND MAPPING IN TRANSITION TO GEOINFORMATION

Geodesy

Geodesy, according to F.R. Helmert (1880), is the science of measurement and mapping of the earth's surface. This involves, first, the determination of a reference surface onto which details of mapping can be fixed.

In ancient Greece (Homer, 800 BC) the earth's surface was believed to be a disk surrounded by the oceans. But, not long after, Pythagoras (550 BC) and Aristotle (350 BC) postulated that the earth was a sphere. The first attempt to measure the dimensions of a spherical earth was made by Eratosthenes, a Greek resident of Alexandria, around 200 BC. At Syene (today's Assuan) located at the Tropic of Cancer at a latitude of 23.5° the sun reflected from a deep well on June 21, while it would not do so in Alexandria at a latitude of 31.1°. Eratosthenes measured the distance between the two cities along the meridian by cart wheel computing the earth's spherical radius as 5909 km. Meridional arcs were later also measured in China (AD 725) and in the caliphate of Baghdad (AD 827).

Until the Renaissance, Christianity insisted on a geocentric concept, and the determination of the earth's shape was not considered important. In the Netherlands, Willebrord Snellius resumed the ancient ideas about measuring the dimensions of a spherical earth using a meridional arc, which he measured by the new concept of triangulation, in which long distances were derived by trigonometry from angular measurements in triangles. The scale was derived from one accurately surveyed small triangle side, which was measured as a base by tape.

The astronomers of the Renaissance—Copernicus (1500), Kepler (1600), and Galileo (1600)—along with the gravitational theories of Newton (1700) postulated that the earth's figure must be an ellipsoid, and that its flattening could be determined by two meridional arcs at high and low latitude. Although the first verification in France (1683–1718) failed due to measurement errors, the measurement of meridional arcs in Lapland and Peru (1736–1737) verified an ellipsoidal shape of the earth. Distances on the ellipsoid could consequently be determined by the astronomical observations of latitude and longitude at the respective points on the earth's surface.

Laplace (1802), C.F. Gauss (1828), and F.W. Bessel (1837), however, recognized that astronomic observations were influenced by the local gravity field due to mass irregularities of the earth's crust. This was confirmed by G. Everest, who observed huge deflections of the vertical in the Himalayas. This led to the establishment of local best-fitting reference ellipsoids for positional surveys of individual countries.

In the simplest case latitude and longitude was astronomically observed at a fundamental point, and an astronomical azimuth was measured to a second point in the triangulation network spanning a country. Within the triangulation network, at least one side was measured by distance-measuring devices on the ground. For the determination of a best-fitting ellipsoid, several astronomic observation stations and several baselines were used. The coordinates of all triangulated and monumented points were calculated and least squares adjusted on the reference ellipsoid with chosen dimensions, for example, for half axis major a and for half axis major b or the flattening.

$$f = \frac{a-b}{a}$$

Clarke, 1880	$a = 6378249$ m, $b = 6356515$ m
Bessel	$a = 6377879$ m, $f = 1/298.61$
Hayford	$a = 6378388$ m, $f = 1/297$
Krassovskij	$a = 6378295$ m, $f = 1/298.4$

On the chosen reference ellipsoid, the ellipsoidal latitudes and longitudes were obtained for all points of the first-order triangulation network. This network was subsequently densified to second-, third-, and fourth-order by lower-order triangulation.

The survey accuracy of these triangulation networks of first to fourth order was relatively high, depending on the observational practices, but discrepancies between best-fitting ellipsoids of neighboring countries were in the order of tens of meters.

For the purpose of mapping, the ellipsoidal coordinates were projected into projection coordinates. Due to the nature of mapping in which local angular distortions cannot be tolerated, conformal projections are chosen:

- For circular countries (e.g., the Netherlands, the province of New Brunswick in Canada), the stereographic projection.
- For N–S elongated countries, the 3° transverse Mercator projection tangent to a meridian, every 3 degrees. Due to its first use by C.F. Gauss and its practical introduction by Krüger, the projection is called the Gauss–Krüger projection. It is applied for meridians 3 degrees apart in longitude in several strips. The projection found wide use for the mapping of Germany, South Africa, and many countries worldwide.
- For E–W elongated countries (e.g., France), the Lambert conic conformal projection was applied to several parallels.
- The Lambert conic conformal projection may also be obliquely applied, for example, in Switzerland.

- For worldwide mapping, mainly for military mapping requirements, the Universal Transverse Mercator (UTM) projection (a 6° transverse Mercator projection) is applied. The formulation is the same as for the Gauss–Krüger projection, with the exception that the principal meridian (here every 6 degrees) has a scale factor of 0.9996 rather than 1 used for the Gauss–Krüger projection.

Since the earth's gravity field influences the flow of water on the earth's surface, ellipsoidal coordinates without appropriate reductions cannot be used for practical height determinations. Height reference systems are therefore separate from position reference systems based on reference ellipsoids.

An ideal reference surface would be the equipotential surface of the resting oceans, called the "geoid." Due to earth tides influenced by the moon and planets, ocean currents, and winds influenced by climate and meteorology, this surface is never resting. For this reason, the various countries engaged in mapping systems have created their own vertical reference systems by observing mean sea level tides at tidal benchmarks. Spirit leveling extended the elevations in level loops of first order over the mapping area of a country to monumented benchmarks. These level loop observations, corrected by at least normal gravity, could be densified by lower-order leveling to the second, third, and fourth orders. As is the case for positions, differences of several meters in height values may be the result of the different height reference systems of different countries.

The different reference systems for position and height still used for mapping in the countries of the world are in transition, changing into a new reference frame of three- or four-dimensional geodesy. This has become possible through the introduction of the U.S. Navy NAVSTAR Global Positioning Systems (GPS) in the 1980s. It now consists of 24 orbiting satellites at an altitude of 20200 km. These orbit at an inclination of 55° for 12 hours, allowing a view, in a direct line of sight, of at least four of these satellites from any observation point on the earth's surface for 24 hours of the day.

Each of the satellites transmits timed signals on two carrier waves with 19.05 cm and 24.45 cm wavelengths. The carrier waves are modulated with codes containing the particular satellite's ephemeris data with its corrections. The U.S. Defense Department has access to the precise P-code suitable for real-time military operations. Civilian users can utilize the less precise C/A code carried by the 19.05 cm carrier wave.

When three satellites with known orbital positions transmit signals to a ground receiver, the observed distances permit an intersection of 3D coordinates on the earth's surface. Since the satellite clocks are not synchronized, an additional space distance from a fourth satellite is required for 3D positioning.

The calculations are based on an earth mass centered reference ellipsoid determined by an observation network by the U.S. Department of Defense, the World Geodetic System 1984 (WGS 84), with the following dimensions:

$a = 6378137$ m

$f = 1/298.257223563$

Local reference ellipsoids used in the various countries differ in coordinate positions by several 100 m with WGS 84 coordinates.

P-code observations may be used in real time to accuracies in the decimeter range. C/A codes are capable of determining positions at the 5 m level unless the satellite clock signals are artificially disturbed by the military satellite system operators, as was the case during the 1990 to 2000 period. This disturbance was called the selective availability (SSA). It deteriorated the C/A code signals to 100 m accuracies in position and to 150 m in height.

To overcome this lack of dependability, more elaborate receivers were developed in the civilian market, which observed the phases of the carrier waves, using the C/A codes only to obtain approximate spatial distances and to eliminate ambiguities when using the phase measurements. The principle of measurement at a mobile rover station thus became that of relative positioning with respect to a permanently operating master reference station.

In the static mode (observing over longer duration periods), positional accuracies in the range of several millimeters could be achieved for distances closer than 10 km. For long distances over several hundreds of kilometers, accuracies in the 1 cm to 2 cm range could be obtained by the simultaneous observation of networks.

Relative observations in networks are able to minimize ionospheric and tropospheric transmission effects. Satellite clock errors may be eliminated using double differences.

Multiple reflection effects may be eliminated by the careful choice of observation points. This has encouraged the international civilian community to establish an International Terrestrial Reference Frame (ITRF) of over 500 permanently observing GPS stations worldwide. The absolute position of an ITRF is combined with the observation of an International Celestial Reference Frame (ICRF), in which the absolute orientation of an ITRF is controlled by stellar observations using radio astronomy (quasars, very long baseline interferometry [VLBI]).

The existence of an ITRF gives the opportunity to monitor changes of plate tectonic movements of the earth's crust. Thus, an ITRF is determined at a specified epoch (e.g., ITRF, 1993, 1997, 2000), in which local plate deformations can be observed that exceed centimeter accuracies.

The existence of an ITRF has encouraged mapping agencies throughout the world to establish new continental control networks and to densify them into national reference systems. In Europe, 36 ITRF stations were selected in 1989 to create the European Terrestrial Reference Frame (ETRF 89). This reference frame served to reobserve national networks with differential GPS, such as the DREF 91 in Germany, which permitted the setting up of a network of permanently observing GPS reception stations, SAPOS, with points about 50 km apart.

Networking solutions, such as those offered by the companies Geo++ and Terrasat, permit the use of transmitted corrections to rover stations observing GPS-phase signals in real-time kinematic mode. These enable positioning in ETRF to 1 cm accuracy in latitude and longitude and to 2 cm accuracy in height at any point of the country where the GPS signals may be observed. Austria, Sweden (SWEPOS), and the Emirate of Dubai have introduced similar systems and now they are common around the globe. Therefore, GPS has become a new tool for detailed local surveys and its updating.

Thus, the problem of providing control for local surveys and mapping operations has been reduced to making coordinate conversions from local reference systems to new geodetic frameworks for data, which have previously been collected.

A detailed coverage of the modern geodetic concept with the mathematical tools has been given in the book *Geodesy* by Wolfgang Torge. A treatise of satellite geodesy is contained in the book *Satellite Geodesy* by Günter Seeber.

Surveying

Although geodesy's main goal was to determine the size and shape of the earth, and to provide control for the orientation of subsequent surveys on the earth's surface, it was the aim of survey technology to provide tools for the positioning of detailed objects.

The first direct distance measurements have been in place since the time of the Babylonians and the Romans. Direct distance observations by tapes and chains, as they have been used in former centuries, have made way to the preference of angular measurement by theodolites, which permitted the use of trigonometry to calculate distances in overdetermined angular triangulation networks. Even for detailed terrestrial topographic surveys, instruments were developed in the early 1900s, the so-called tacheometers, which were able to measure distances rapidly in polar mode by optical means, combining these with directional measurements.

In the 1950s, the direct measurement of distances became possible by the invention of electronic distance-measuring devices. The Tellurometer, developed in South Africa, utilized microwaves as a carrier onto which measurement phases of different wavelengths were modulated. The Swedish geodimeter used light as a carrier and later most manufacturers used infrared carrier waves. These distance-measuring capabilities were soon combined with directional measurement devices, known from theodolites in the form of electronic tacheometers. The automatic coding of directional and distance measurements was perfected in the so-called total stations.

Nowadays it is possible to integrate GPS receivers and PC graphic capabilities in a single instrument, the so-called electronic field book, which can be used effectively for updating vector graphics or for orienting oneself in the field with the help of digitized and displayed images. For the survey of heights, simple route survey devices, such as the barometer, are considered to be too inaccurate according to today's standards.

Leveling is still the prime source of accurate height data, even though spirit level devices have gradually been replaced by instruments with automatic compensators assuring a horizontal line of sight by the force of gravity. The optically observed level rods read by the operator have likewise been automated by digitally coded reading devices to permit more rapid leveling operations.

Nevertheless, ground surveys without the use of GPS are still considered very expensive, and they are thus only suitable for the detailed survey of relatively small areas.

Remote Sensing

Remote sensing can be considered as the identification or survey of objects by indirect means using naturally existing or artificially created force fields. Of most significant impact are systems using force fields of the electromagnetic spectrum that permit the user to directionally separate the reflected energy from the object in images.

The first sensor capable of storing an image, which could be later interpreted, was the photographic emulsion, discovered by Nièpce and Daguerre in 1839. When images were projected through lenses onto the photographic emulsion, the photographic camera became the first practical remote sensing device around 1850.

As early as 1859, photographs taken from balloons were used for military applications in the battle of Solferino in Italy and later during the American Civil War. Only after the invention of the aircraft in 1903 by the Wright brothers did a suitable platform for aerial reconnaissance become of standard use. This was demonstrated in World War I, during which the first aerial survey camera was developed by C. Messter of the Carl Zeiss Company in Germany in 1915. Aerial photographic interpretation was extended into many application fields (e.g., glaciology, forestry, agriculture, archaeology), but during World War II it again became the primary reconnaissance tool on all sides.

In Britain and Germany, development of infrared sensing devices began, and Britain was successful in developing the first radar in the form of the plan position indicator (PPI). Further developments were led by the United States in the postwar years, developing color-infrared film in the 1950s. Other developments went on in side-looking airborne radar (SLAR, SAR).

In the 1960s, remote sensing efforts made use of the first satellite platforms. TIROS was the first meteorological satellite. The lunar landing preparations for the Apollo missions of NASA had a strong remote sensing component, from sensor development through to analysis. When the lunar landing was accomplished, NASA turned its interest toward remote sensing of the earth's surface.

In 1972, the Earth Resources Technology Satellite (ERTS-1), which later was called Landsat 1, became the first remote sensing satellite with a world coverage at 80 m pixels in four spectral visible and near-infrared channels. In subsequent years both spatial and spectral resolution of the first Landsat were improved: Landsat 3, launched in 1982, had six visible and near-infrared channels at 30 m pixels and one thermal channel at 120 m pixels.

Higher spatial resolution was achieved by the French Spot satellites launched since 1986 with panchromatic pixel sizes of 10 m and multispectral resolution at 20 m pixels. The Indian satellites IRS 1C and 1D reached panchromatic pixel sizes of 6 m in 1996. Even higher resolution with photographic cameras was reached from space in the U.S. military Corona program of the 1960s (3 m resolution) and the Russian camera KVR 1000 (2 m resolution) in 1991. Since 1999, the U.S. commercial satellite Ikonos 2 has been in orbit, which produces digital panchromatic images with 1 m pixels on the ground. This was surpassed by the U.S. commercial satellite Quickbird, with 0.6 m pixels on the ground.

In 1978, the first coherent radar satellite Seasat was launched by the United States, but it only had a very short lifetime. In 1991, the European Space Agency (ESA) commenced a radar satellite program ERS 1 and 2, which was followed by the Japanese radar sensor on JERS 1 in 1994, the Canadian Radarsat in 1995, and the Russian Almaz in 1995. The NASA Space Shuttle Radar Topography Mission (SRTM) in the year 2000 carried an American C band radar sensor and a German X band radar sensor by which large portions of the land mass of the earth were imaged at pixel sizes of 25 m to 30 m. Coherent radars are not only of interest due to their all-weather, day and night capability, but due to the possibility of deriving interferograms from adjacent or subsequent images, which permit the derivation of relative elevations at 12 m to 6 m accuracy.

A new development is the construction of hyperspectral sensing devices, which, at lower resolutions, permit the scanning of the earth in more than 1000 narrow spectral bands to identify objects by their spectral signatures.

Multisensoral, multispectral, multitemporal, and in the case of radar, even multipolarization images permit vast possibilities for image analysis in remote sensing, which are not yet fully explored.

PHOTOGRAMMETRY

Photogrammetry is the technology to derive geometric information of objects with the help of image measurements. Although the use of perspective geometry dates back to the Italian Renaissance (Brunneleschi, 1420; Piero della Francesa, 1470; Leonardo da Vinci, 1481; and Dürer, 1525), it needed the invention of photography to develop a usable tool. In 1859, the military photographer Aime Laussedat used terrestrial photographs for the construction of plans for the city of Paris. In 1858, the architect Meydenbauer used terrestrial photographs for the derivation of plans for the cathedral of Wetzlar in Germany, and Sebastian Finsterwalder of Munich used photographs for the survey of glaciers in the Tyrolean Alps in 1889.

Terrestrial and balloon photographs were, however, not suitable for providing a systematic coverage of the earth's surface. This was made possible by the use of aircraft combined with the use of aerial survey cameras during World War I.

Even though Sebastian Finsterwalder made a rigid reconstruction of a pair of balloon photographs of the area of Gars am Inn, Germany, in 1899, using mathematical calculations of many points measured in the photos, this analytical approach, without the availability of computers, was too time consuming. Therefore, analogue instrumentation with stereo measurement was developed, which permitted fast optical or mechanical reconstruction of the photographic rays determining an object point. In the 1920s, German, Italian, French, and Swiss designers developed a great variety of photogrammetric plotting instruments. These were used during World War II to meet the rapidly increased mapping demands of the nations involved in the war.

After World War II, modifications and simplifications of the optical and mechanical plotting instruments developed in Switzerland, Britain, France, Italy, the United States, and the USSR played an important role in meeting basic mapping demands all over the globe. In order to diminish control requirements for mapping, aerial triangulation was developed as a method to interpolate a scarce control distribution down to the requirements of a single stereo model.

The introduction of the computer into photogrammetry in the late 1950s not only permitted part-automation of the tasks of aerial triangulation and stereo restitution, but also aided in the increased accuracy and reliability of the restitution process.

Computer developments in the 1970s and 1980s, with increased speed and storage, finally permitted users to treat aerial images in digital form after they have been scanned by raster scanners. Digital photogrammetry combined with image processing techniques finally became the new tool to partially or fully automate point measurement, coordinate transformation, image matching for the derivation of the third dimension, and for differential image rectification to create orthoimages with the geometry corresponding to a map. This technology is not only suitable to be applied to aerial photographs, but it can likewise also be used, with slight modifications, for terrestrial images or for digital images of satellite scanners.

The development of laser scanners for use in aircraft is relatively new. Successful systems have been developed to measure the distance between scanners and the terrain with accuracies in the range of 10 cm. The advantage of the technique is that it can be used in urban areas and in forests obtaining multilevels of terrain information (e.g., terrain versus rooftops of buildings, terrain versus treetops). A proper restitution technology, however, requires that the sensor position and the sensor orientation is accurately known.

Attempts to measure sensor position and orientation date back to the 1920s, when instrumentation for the measurement of some orientation components was made possible by devices such as the statoscope, in which barometric pressure changes were used to derive elevation differences between exposure stations. Other devices, such as the horizon camera, aimed at determining the sensor orientation. Today, in-flight GPS permits the establishment of sensor positions in real time with 15 cm accuracy. Inertial measuring units permit the determination of sensor inclination with a correspondingly high accuracy. The aim is to solve the control densification problem without the need for aerial triangulation. A historical review of the developments in photogrammetry is given in Konecny (1996).

Geographic Information Systems (GIS)

Geographic information systems arose from activities in four different fields:

- Cartography, which attempted to automate the manually dependent map-making process by substituting the drawing work by vector digitization.
- Computer graphics, which had many applications of digital vector data apart from cartography, particularly in the design of buildings, machines, and facilities.
- Databases, which created a general mathematical structure according to which the problems of computer graphics and computer cartography could be handled.
- Remote sensing, which created immense amounts of digital image data in need of geocoded rectification and analysis (see Figure 1.1).

The first theoretical and exploratory attempts for a GIS design started in the 1960s, namely by:

- R.F. Tomlinson, who in 1968, created the first Canada Geographic Information System for the Agricultural Research Development Agency (ARDA) of Canada.

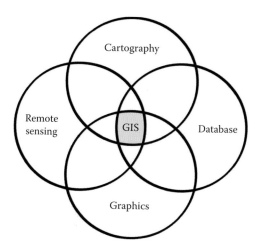

Figure 1.1   Interrelationship between GIS disciplines.

- The Experimental Cartography Unit under Bickmore in the United Kingdom, which has attempted to automate cartography since 1963.
- The Harvard Laboratory for Computer Graphics, which has laid the theoretical foundations for successful industrial GIS developments since 1964, through its creation of Symap (e.g., Jack Dangermond for ESRI and D.F. Sinton for Intergraph).
- The Swedish attempts to establish a Land Information System (LIS) for the district of Upsala.

These first personal initiatives were followed by the takeover of these ideas by governmental administrations:

- U.S. Bureau of the Census, from 1967, utilizing DIME files.
- U.S. Geological Survey, creating Digital Line Graphs.
- Attempts of the German states and their surveys and mapping coordination body, ADV, creating the ATKIS concept.

The successful introduction of the technology without subsequent strong industrial development would not have been possible. The leaders in the field from 1969 were:

- ESRI under Jack Dangermond, in the United States.
- Intergraph under Jim Meadlock, in the United States.
- Siemens in Germany.

As the developments in computing speed and computer storage grew rapidly in the 1980s and 1990s, it became obvious that an integration of industrial efforts for user requirements was mandatory. This led to the creation of the Open GIS Consortium (OGIS), which, along with the International Standards Organization (ISO) and the European Standards Organization (CEN), led to the transferability of data created on different systems, to the integration of vector and raster data, to the establishment of metadata bases and to the transmission of data on the Internet or Intranet.

It was soon realized that data created and analyzed for project purposes only was a duplication of effort, and that the tendency went into the creation of base data, which had to be continuously updated not only to provide an inventory of geospatial information, but to permit its cross-referenced analysis and to integrate the effort into a management system for sustainable development. A review of the historical GIS developments is given in the book, *Geographical Information Systems*, by Longley, Goodchild, Maguire, and Rhind.

CURRENT STATUS OF MAPPING IN THE WORLD

The United Nations Secretariat has tried to monitor existing base map data for different countries and continents at different scale ranges between the 1960s and the 1980s (see Table 1.1).

A near global coverage only existed at the scale range of 1:200000 or 1:250000. At 1:50000, about two-thirds of the land area is covered and at 1:25000 about one-third. The coverage at these scales is in analogue form, but is subject to progress in vectorization or at least in raster scanning. No data surveys exist at large scales.

At the United Nations Cartographic Conference in Bangkok in 1990, a survey on the annual update rates of these maps was presented, as shown in Table 1.2.

The conclusion of these surveys is that the update rate for the 1:50000 map is only 2.3%. This means that the average existing 1:50000 map of the countries of the world is about 44 years old, and that the average existing 1:25000 map is 20 years old.

TABLE 1.1   STATUS OF WORLD MAPPING (1986)

Scale Range	1:25000	1:50000	1:100000	1:200000
Africa	2.9%	41.4%	21.7%	89.1%
Asia	15.2%	84%	56.4%	100%
Australia and Oceania	18.3%	24.3%	54.4%	100%
Europe	86.9%	96.2%	87.5%	90.9%
Former USSR	100%	100%	100%	100%
North America	54.1%	77.7%	37.3%	99.2%
South America	7%	33%	57.9%	84.4%
World	33.5%	65.6%	55.7%	95.1%

TABLE 1.2   UPDATE RATES OF WORLD MAPPING (1986)

Scale Range	1:25000	1:50000	1:100000	1:200000
Africa	1.7%	2.2%	3.6%	1.4%
Asia	4.0%	2.7%	0%	1.9%
Australia and Oceania	0%	0.8%	0%	0.3%
Europe	6.6%	5.7%	7.0%	7.5%
Former USSR	0%	0%	0%	0%
North America	4.0%	2.7%	0%	6.5%
South America	0%	0.1%	0%	0.3%
World	5.0%	2.3%	0.7%	3.4%

While ground survey methods applied in the 19th century have been able to create national coverages in Europe within a century, aerial photogrammetry applied in the 20th century was able to provide mapping coverages in other continents, at least in priority areas. But technological, financial, or organizational constraints have not been able to provide an updated basic mapping coverage needed for sustainable development. Therefore, assistance by new geoinformation technologies is required to achieve the necessary progress.

INTEGRATION OF GEOINFORMATION TECHNOLOGIES

Figure 1.2 outlines the differences between the classical spatial information system approach and the new geospatial information concept. In the past, the disciplines of geodesy, photogrammetry, and cartography worked in an independent fashion to provide printable maps. In the new concept, GPS, remote sensing, digital photogrammetry, and GIS are able not only to produce printable maps, but to display raster and vector images on a computer screen and to analyze them in an interdisciplinary manner for the purposes of society (see Figure 1.2).

UNITED NATIONS INITIATIVE ON GLOBAL GEOSPATIAL INFORMATION MANAGEMENT

In 2010, the United Nations Secretariat created a new initiative, the "global geographical information management" (UN-GGIM) under the UN Statistics Division. Its aim is to ensure the underavailability of up-to-date geodata on a global, regional, and national level. Based on an initiative by the International Society for Photogrammetry and Remote Sensing (ISPRS) UN-GGIM, backed by a United Nations Economic and Social Council (UN ECOSOC) resolution, decided to resume the former UN Secretariat efforts by conducting a new study to

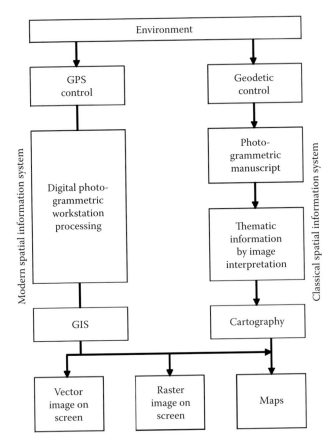

Figure 1.2 Classical and modern geospatial information systems.

which the 193 UN member countries were invited. The data received from this inquiry found that during the past 20 to 30 years mapping has become digital. Even if only half of the UN member countries responded, the largest countries were among them. For 89 countries, 22 had full country coverage at 1:1000, 40 at 1:5000, 46 at 1:25000, and 70 at 1:50000. In 6% of those countries, these maps were free of charge and in 71% they were available for sale. In 23% of the countries, they were restricted from the public.

Most of the countries (83%) had updating programs based on aerial photography as an in-house effect by the national mapping agencies. In 60% of the countries, satellite imagery was also used for more rapid preliminary updating.

A national coverage of cadastral maps existed only in 69% of the countries.

In 93% of the countries, maps were available on digital media. Nearly half of the countries (48%) distributed their maps via the Internet. Of interest were the regional differences. For Europe the base scales were in general between 1:1000, 1:5000, and 1:25000. In Africa, the base scale was 1:50000. In Europe, 93% of the countries had regular aerial photography update programs, but in Africa only 56%.

In Europe, 82% of the countries had a country coverage of cadastral maps, but in the Americas only 45%.

In the United States, the basic topographic map is at the scale of 1:24000. It is updated every 3 years. A high accuracy digital elevation model based on airborne laser scanning is available for the Eastern third of the United States. In Alaska, where the 1:24000 coverage does not exist, mapping is done from radar imaging, and the digital elevation model (DEM) is generated from interferometric SAR.

The international map vendors, East View Geospatial in Minneapolis, Minnesota, and ILH in Stuttgart, Germany, offer services to provide map coverages of parts of the world, where the coverages is restricted to the public. They indicate that map coverages, which are eventually outdated, are still available from military sources in the United States from NATO and from the Russian Federation at the scale of 1:50000 (see Figure 1.3). It is interesting that mapping agencies in most countries now deliver their products in vector or raster form by servers (see Figure 1.4). It is also a fact that many countries do not have a full cadastral map coverage (see Figure 1.5).

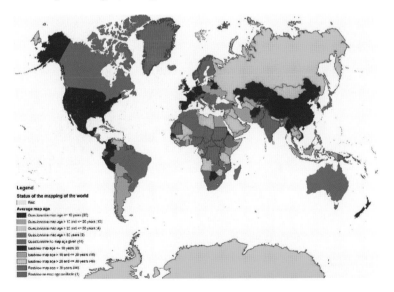

Figure 1.3 Status and age of mapping in the world—2013. Note: Dark green = less than 10 years; Light green = less than 20 years; Yellow = less than 30 years; Red = more than 30 years; Gray = information not available. (From Konecny, G., "Status of Mapping in the World," ISPRS Publication to UN-GGIM3 Conference, Cambridge, England, 2013. With permission.)

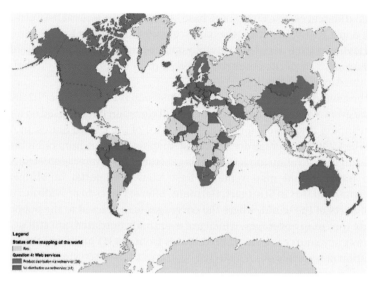

Figure 1.4   Delivery of maps by servers. Note: Green = delivery of maps by servers; Red = delivery of maps by other media. (From Konecny, G., "Status of Mapping in the World," ISPRS Publication to UN-GGIM3 Conference, Cambridge, England, 2013. With permission.)

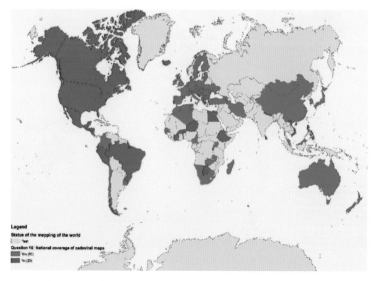

Figure 1.5   Coverage of cadastral systems. Note: Green = cadastral map coverage exists; Red = cadastral map coverage does not exist. (From Konecny, G., "Status of Mapping in the World," ISPRS Publication to UN-GGIM3 Conference, Cambridge, England, 2013. With permission.)

Of further interest are the activities that high-resolution satellite imagery provides, such as Digital Globe for GeoEye and WorldView images.

According to the subvendor for satellite imagery, Scanex, in the Russian Federation 61% of the territory is covered by GeoEye 1 and World View 2 imagery at 0, 5 m ground sample distance. Georeferenced imagery therefore is useful as a map substitute.

Another major global imaging and mapping effort is carried out by Google Earth and Google maps and by the Microsoft Bing Maps program. Bing Maps has covered the United States as well as western and central Europe at 30 cm ground sample distances (GSD) and their urban areas at 15 cm GSD by aerial orthoimagery for their mapping.

Mapping of partial features is being carried out by Navteq and TomTom for many countries of the world for road navigations systems. Although such products often do not meet the standard mapping specifications set by national governmental map agencies, they fulfill a public demand for urgent and updated geographical information.

Chapter 2

Remote Sensing

Remote sensing is a method of obtaining information from distant objects without direct contact. This is possible due to the existence or the generation of force fields between the sensing device and the sensed object. Usable force fields are mechanical waves in solid matter (seismology) or in liquids (sound waves). But the principal force field used in remote sensing is that of electromagnetic energy, as characterized by the Maxwell equations. The emission of electromagnetic waves is suitable for directional separation. Thus, images of the radiation incident on a sensor may be generated and analyzed.

The remote sensing principle using waves of the electromagnetic spectrum is illustrated in Figure 2.1. The energy radiates from an energy source. A passive (naturally available) energy source is the sun. An active energy source may be a lamp, a laser, or a microwave transmitter with its antenna. The radiation propagates through a vacuum with the speed of light, c, at about 300,000 km/second. It reaches an object, where it interacts with the matter of this object. Part of the energy is reflected toward the sensor. At the sensor, which is carried on a platform, the intensity of the incoming radiation is quantized and stored. The stored energy values are transformed into images, which may be subjected to image processing techniques before they are analyzed to obtain object information.

ELECTROMAGNETIC RADIATION

Basic Laws

Electromagnetic energy is radiated by any body having a temperature higher than –273°C (or 0 K), the absolute zero temperature. Such a body radiates energy in all frequencies. The relation between frequency, ν, and wavelength, λ, is expressible as

$$\lambda = \frac{c}{\nu}$$

with λ expressed in meters and frequency in cycles per seconds (i.e., Hertz).

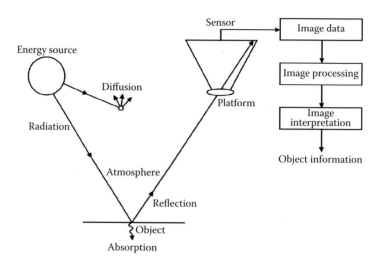

Figure 2.1   Principles of remote sensing.

The amount of radiation for a particular wavelength over an interval, $\Delta\lambda$, is a function of the absolute temperature of the body in kelvins and is expressed by Planck's distribution law

$$L_\lambda(T) = \frac{2h \cdot c^2}{\lambda^5} \cdot \frac{1}{e^{h \cdot c/\lambda \cdot k \cdot T} - 1}$$

where
$L_\lambda(T)$ = spectral radiance in W/m²·sr·µm
$k = 1.38047 \cdot 10^{-23}$ [w·s·K^{-1}], the Boltzmann constant
$h = 6.6252 \cdot 10^{-34}$ [J·s], the Planck's constant
$c = 2.997925 \cdot$cm⁸/s

This is based upon the observation that energy is emitted in energy quantums, Q_p [J = W·s]. Wien's law permits us to determine the radiation maximum at a particular wavelength, λ_{max}, depending on the body temperature by differentiating $L_\lambda(T)$ with respect to λ:

$$\lambda_{max} = \frac{0.002898}{T} \cdot \frac{[mL]}{[K]}$$

Since the surface temperature of the sun is about 6000 K, this means that the maximum radiance from solar energy will be generated at a wavelength of 480 nm, which corresponds to green light. The earth, with its temperature

TABLE 2.1   RADIOMETRIC QUANTITIES

	Name	Symbol	Relation	Dimension	
Radiator	Radiant energy	Q	–	Joule	$J = W \cdot s$
	Radiant flux	Φ	$\Phi = \dfrac{Q}{t}$	Watt	$W = J/s$
	Radiant intensity	I	$I = \dfrac{Q}{t \cdot \omega_1}$	—	W/sr
	Radiance	L	$L = \dfrac{Q}{t \cdot A_1 \omega_1}$	—	$W/sr \cdot m^2$
Emitted object	Radiant emittance	E	$E = \dfrac{\Phi}{A_1}$	—	W/m^2
Received signal	Irradiance exposure	E'	$E' = \dfrac{\Phi}{A_2}$	—	W/m^2
		H	$H = E' \cdot t$	—	$W \cdot s/m^2$

Notes: ω_1, space angle of radiation in sterad (sr); A_1, radiating surface in square meters (m^2); A_2, irradiated surface in square meters (m^2).

between 273 and 300 K (0°C and 27°C) has its radiance maximum in the thermal range of the spectrum (8 μm to 14 μm).

Radiometric Quantities

The dimensions of radiometric quantities used in absolute remote sensing are shown in Table 2.1.

Electromagnetic Spectrum

The characteristics of electromagnetic energy in the electromagnetic spectrum are shown in Table 2.2.

ENERGY–MATTER INTERACTION

Atmospheric Transmission

Electromagnetic transmission through the atmosphere slows the wave propagation depending on the transmission coefficient, n:

$$\lambda = \frac{c}{v \cdot n}$$

TABLE 2.2   ELECTROMAGNETIC SPECTRUM

Radiation Type	Wavelength	Frequency	Transmission	Use	Detector
Cosmic rays	10^{-13} to 10^{-16}	4.7×10^{21} Hz to 3×10^{24} Hz	Outer space	—	Ionization detector
γ-rays	10^{-4} nm to 0.4 mm (nm = 10^{-9} m)	8×10^{16} to 4.7×10^{21} Hz	Limited through atmosphere	Radioactivity	Ionization detector
X-rays	0.4–10 nm	3×10^{16} to 8×10^{16} Hz	Only at close range outer space	Close range	Phosphorus
Ultraviolet light	10–380 nm	7.9×10^{14} to 3×10^{16} Hz	Weak through atmosphere	—	Phosphorus
Visible light	380–780 nm	3.8×10^{14} to 7.9×10^{14} Hz	Well through atmosphere	Passive remote sensing, vision	Photography, photodiode
Near infrared	780 nm–1 μm (m = 10^{-6} km)	3.0×10^{24} to 3.8×10^{14} Hz	Well through atmosphere	Passive remote sensing	Photography, photodiode
Medium infrared	1–8 μm	3.7×10^{13} to 3.0×10^{14} Hz	In windows through atmosphere	Passive remote sensing	Quantum detector
Thermal infrared	8 μm–1 mm (mm = 10^{-3})	3×10^{11} to 3.7×10^{13} Hz	In windows through atmosphere, day and night	Passive remote sensing	Quantum detector
Thermal infrared	8 μm–14 μm (to 1 mm) (mm = 10^{-3})	3×10^{11} to 3.7×10^{13} Hz	In windows through atmosphere, day and night	Passive remote sensing	Quantum detector

	1 mm–1 m	300 MHz to 300 GHz (MHz = 10^6 Hz, GHz = 10^9 Hz)	Day and night through clouds	Active remote sensing	Antenna
Microwaves	1 mm–1 m	300 MHz to 300 GHz (MHz = 10^6 Hz, GHz = 10^9 Hz)	Day and night through clouds	Active remote sensing	Antenna
FM radio	1–10 m	30–300 MHz	Direct visibility	TV, broadcast	Antenna
Short-wave radio	10–100 m	3–30 MHz	Worldwide	Broadcast	Antenna
Medium-wave broadcast	182 m–1 km	300–1650 KHz (KHz = 10^3 Hz)	Regional	Broadcast	Antenna
Long-wave broadcast	1–10 km	30–300 KHz	Worldwide	Broadcast	Antenna
Sound transmission	15–600 km	50 Hz–20 KHz	Cable	Telephone	Cable
AC	6000 km	50 Hz	Cable	Energy transmission	Cable

Furthermore, part of the energy is absorbed when the energy quantums hit the molecules and atoms of the atmospheric gases.

Another part is directionally reflected into diffused energy causing scattering, according to Rayleigh.

$$I_{transmitted} = I_{original} \cdot \exp(-k{\cdot}r)$$

where
   k = extinction coefficient
   r = distance passed through
   k is proportional to λ^{-4}

In the infrared ranges of the spectrum, absorption takes place by gases contained in the atmosphere, so that transmission is possible only in atmospheric windows, as shown in Figure 2.2.

Energy Interaction at the Object

Interaction of incident electromagnetic energy with matter depends on the molecular and atomic structure of the object. Energy may be directionally reflected, scattered, transmitted, or absorbed. The process is caused by the interaction of a photon with an electron located in a shell of an atom, which results in the excitation of the electron from the shell. The ratio between the reflected (in all directions), transmitted, and absorbed fluxes or radiances and the incoming radiation is described as:

   Reflection coefficient, ρ_λ
   Transmission coefficient, τ_λ
   Absorption coefficient, α_λ

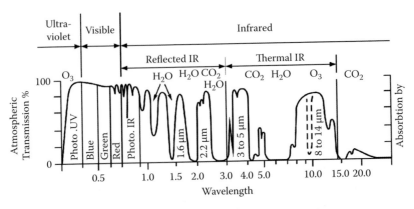

Figure 2.2 Atmospheric windows.

Their sum is equal to 1:

$$\rho_\lambda = \frac{\phi_{\lambda \; \text{reflected}}}{\phi_{\lambda \; \text{incoming}}}$$

$$\tau_\lambda = \frac{\phi_{\lambda \; \text{transmitted}}}{\phi_{\lambda \; \text{incoming}}}$$

$$\rho_\lambda = \frac{\phi_{\lambda \; \text{absorbed}}}{\phi_{\lambda \; \text{incoming}}}$$

$$\rho_\lambda + \tau_\lambda + \alpha_\lambda = 1$$

Other than through atmospheric particles, transmission is possible, for example, for water. Nontransparent solid bodies have a transmission coefficient of 0.

The incoming energy cannot get lost. Absorption is a process in which higher frequency energy (e.g., light) is converted to lower frequency energy (e.g., heat).

The reflection coefficient of an object is of crucial importance for remote sensing. It varies for different spectral ranges for a particular object. It is characterized by an angle dependent function, the so-called radiometric function:

$$\rho_\lambda(\in_1, \in_2, \in_3) = \rho_{\lambda 0} f(\in_1, \in_2, \in_3) = \rho_\lambda \text{ (solar position)}$$

in which $\rho_{\lambda 0}$ is the normal reflection coefficient, valid for an object illuminated and reflected in the same direction. $\in_1$ is the spatial angle between the direction of illumination and the surface normal. $\in_2$ is the spatial angle between the direction of the sensor and the surface normal. $\in_3$ is the spatial angle between the direction of illumination and the direction to the sensor.

The normal reflection coefficients vary for different object types, for example, for green light:

Coniferous forest, 1%
Water, 3%
Meadow, 7%
Road, 8%
Deciduous forest, 18%
Sand, 25%
Limestone, 60%
New snow, 78%

The spectral reflectance varies with each object type, as shown in Figure 2.3 for soil and vegetation.

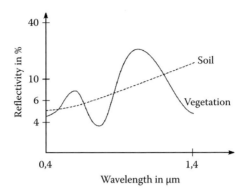

Figure 2.3   Spectral reflectance.

Interaction of Energy in Plants

Whereas the reflection on smooth object surfaces is simple, the interaction of energy in plants is more complicated. A plant leaf consists of at least three layers (Figure 2.4):

- Transparent epidermis
- A palisade type mesophyll, which reflects green light and absorbs red, due to its chlorophyll content
- A spongy type mesophyll that reflects near infrared

Energy Arriving at the Sensor

The radiant density flux arriving at a sensor is composed of the following terms: the radiant flux of solar radiation, E_λ, is diminished by the solar altitude ($90° - \in_1$) and the transmission coefficient of the atmosphere $\tau_{\lambda_{\in_1}}$ for the incident ray to result in the radiant flux of the object: $E_{\in_1}$:

$$E_{\in_1} = \cos \in_1 \int_{\lambda_1}^{\lambda_2} E_\lambda \cdot \tau_{\lambda_{\in_1}} \cdot d\lambda$$

The reflection at the object is dependent on the radiometric function. In case the object is a Lambert reflector, reflecting all incident energy into the half sphere with equal intensity, $f(\in_1, \in_2, \in_3)$ becomes equal to 1.

The radiant flux arriving at the sensor then becomes

$$E_{\in_2} = \frac{\cos \in_1}{\pi} \frac{1}{} \int_{\lambda_1}^{\lambda_2} E_\lambda \cdot \tau_{\lambda_{\in_1}} \cdot \tau_{\lambda_{\in_2}} \cdot \rho_\lambda \cdot d\lambda$$

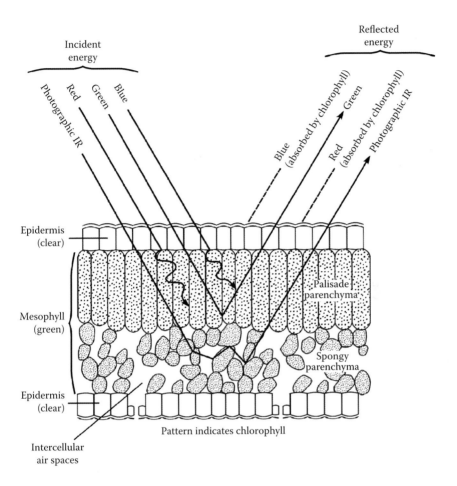

Figure 2.4   Cross-section of a leaf.

with $\tau_{\lambda_{\epsilon 2}}$ as the transmission coefficient of the atmosphere between object and sensor, and the reflection coefficient, ρ_λ, equal in all directions of reflection.

The actual radiant flux arriving at the sensor, as far as its intensity is concerned, is still augmented, however, by the diffused light of the atmosphere $\rho_{\epsilon_1\lambda}$, which amounts to about 3%. It is strongly wavelength dependent.

Thus, the incident radial flux, E, at the sensor becomes:

$$E = \frac{1}{\pi} \int_{\lambda_1}^{\lambda_2} E_\lambda (\tau_{\lambda_{\epsilon_1}} \cdot \tau_{\lambda_{\epsilon_2}} \cdot \rho_\lambda \cdot \cos \epsilon_1 + \rho_{\epsilon_1\lambda}) d\lambda$$

E_λ can be derived from solar parameters. The transmission coefficients of the atmosphere can be determined from radiometric measurements of the solar illumination with radiometers directed at the sun and the total illumination of the half sphere, measurable by an Ulbricht sphere in front of the radiometer. With $\in_1$ known and $\rho_{\in_{1\lambda}}$ estimated for the diffused light, remote sensing can be treated as a procedure of absolute radiometry.

To use this absolute methodology is, however, impractical because of the need to measure atmospheric transmission parameters and because of assumptions for the radiometric function and the scattered energy.

Thus, remote sensing generally restricts itself to a relative comparison of directionally separable radiant fluxes or radiances of adjacent objects A and B

$$\frac{E_A}{E_B} = \frac{L_A}{L_B} = \frac{\rho_A}{\rho_B}$$

as E_λ, $\tau_{\lambda_{\in_1}}$, $\cos \in_1$, and $d'(x,y)$ are nearly equal in this case.

SENSOR COMPONENTS

The classical remote sensing device is photography.

Optical Imaging

The first historical device to create an image of a scene was the pinhole camera. It contains a small circular hole, which can be considered as the projection center. Light rays passing through this hole may be imaged in an image plane at an arbitrary distance.

The disadvantage of the pinhole camera is that the incident radiation in the image plane is too weak for practical purposes. Optical lenses permit the collection of more radiation. To be imaged the lenses must, however, focus the image according to the condition

$$\frac{1}{a} + \frac{1}{b} = \frac{1}{f}$$

in which a is the distance between the projection center and object point, b is the distance between the projection center and image point, and f is the focal length of the spherically shaped lens.

Simple lenses suffer a number of sharpness restrictions, such as

- Chromatic aberration, which focuses light of different wavelengths in separate image planes

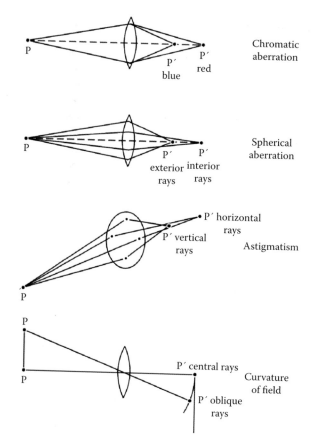

P — Chromatic aberration
P′ blue   P′ red

P — Spherical aberration
P′ exterior rays   P′ interior rays

P′ horizontal rays
P′ vertical rays
Astigmatism
P

P — Curvature of field
P′ central rays
P′ oblique rays

Figure 2.5 Lens errors.

- Spherical aberrations, which collect the inner rays of the lens in different planes with respect to the outer rays of the lens
- Astigmatism, which causes the rays passing vertically through the lens to be collected in a different plane than for the horizontal rays
- Curvature of the field causes the outer rays and the inner rays passing through the lens not in a plane but on a spherical surface (see Figure 2.5)

Lens manufacturers have therefore attempted to combine lenses of different refractive indices, correcting the sharpness deficiencies of simple lenses.

The amount of energy arriving in the image plane through a lens system with the diaphragm, d, is a function of the angle α between its optical axis and the imaged point. This function is influenced by the longer path of the ray, the transmission coefficient of the lens system, and the diameter. For a single lens, the irradiance diminishes with $\cos^4\alpha$. Objectives are able to reduce the light fall off to a function $\cos^{2.5}\alpha$.

The exposure, H, at a point on the optical axis of the image plane thus becomes

$$H = E \cdot \left(\frac{d}{f}\right)^2 \cdot \frac{\pi}{4} \cdot \tau_0 \cdot t$$

with E being the irradiant radial flux at the sensor, d the diaphragm of the optics, f the focal length, t the exposure time, and τ_0 the transmission coefficient of the objective in the direction of the object, which is α dependent.

Photographic Process

The optically produced image in the image plane can be recovered by the photographic process on a film. A layer of film consists of silver halides embedded in gelatin forming the emulsion. When light falls on the emulsion, the silver halide particles are in part chemically converted into metallic silver. The amount of this conversion is proportional to the irradiance.

During the wet development process in the laboratory by methylhydrochinon, the conversion process is continued. A second wet process, the fixation, stops the conversion and dissolves the remaining silver halides. After a washing process, the metallic silver becomes visible. Thereafter, the film is subjected to a drying process in which a film negative that can be copied is produced. It may be copied by illumination onto a copying film, which is subjected to the same process of development, fixation, washing, and drying to produce a diapositive.

The gray level of the produced metallic silver on negative or diapositive corresponds to a logarithmic function of the irradiance. This is shown in Figure 2.6 as the so-called D–log E curve, which should better be called the D–log H curve linking photographic density to irradiance.

The D–log H curve is characteristic for a particular film type. Exposure of the entire irradiance range of the scene should be chosen within the linear range of the D–log H curve. In this range, the slope of the curve is equal to the gradation γ:

$$tg\alpha = \frac{\Delta D}{\Delta \log H} = \gamma$$

The gradation of the film can be influenced by the development process.

$\gamma < 1$ is called a soft development
$\gamma = 1$ is a normal development
$\gamma >$ is a hard development

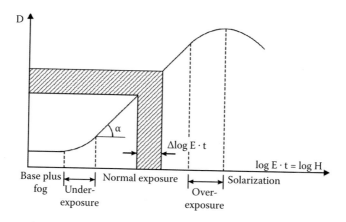

Figure 2.6   D–log H curve.

γ can be changed by the choice of the photographic material, by the developer type, by the temperature of the developer, and by the duration of the development process.

If the development process is fixed as a standard for the type of developer, temperature, and the duration of development, then the sensitivity of the film material can be defined (see Figure 2.7).

A point A on the density curve is reached when the density difference ΔD above fog is 0.1. For this point, the sensitivity in ASA or ISO is

$$\frac{0.8}{H_A}$$

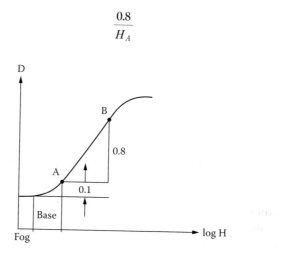

Figure 2.7   Photographic sensitivity.

and the sensitivity in DIN is:

$$10 \cdot \log \frac{1}{H_A}$$

The development process is fixed as a standard so that an irradiance difference $\Delta \log H_{AB} = 1.30$ corresponds to a density difference $\Delta D_{AB} = 0.8$. Silver halides are most sensitive to blue light, which is heavily influenced by scattering in the atmosphere. If this scattering effect is to be minimized, imaging through a yellow filter is mandatory.

The addition of optical dyes permits us to extend the sensitivity of a film into the near-infrared region, not visible to the eye, in which vegetation reflects heavily. Thus, special film types have become available for the visible spectrum or for the spectral range from green to infrared, filtering out the blue light (see Figure 2.8).

The gray value of the negative or diapositive is determined by the transparency of the developed emulsion. Transparency is the ratio between the light flux passing the emulsion ϕ and the incoming light flux ϕ_0:

$$\tau = \frac{\phi}{\phi_0}$$

$1/\tau$ is called the opacity; density D is defined as:

$$D = \log \frac{1}{\tau}$$

Color and Color Photography

As human vision has the ability to distinguish different frequency components of the radiances of objects in the form of color, a definition of the color system has been necessary. The Commission Internationale de l'Eclairage (CIE) has defined the three principal colors for the following wavelengths:

Blue = 435.8 nm
Green = 546.1 nm
Red = 700.0 nm

Each perceived color corresponds to an addition of the three principal color components. All colors can be represented in the CIE chromaticity diagram shown in Figure 2.9.

If the colors red, green, and blue are equally represented, their mixing appears to the eye as white at a particular intensity (brightness). On the outside curve of the diagram are pure colors corresponding to a particular frequency between 380 nm and 770 nm. These represent the hue (dominant wavelength) of the light received. Inside the diagram a particular light, coming from a mix of frequencies, is represented by the saturation (purity) of a color.

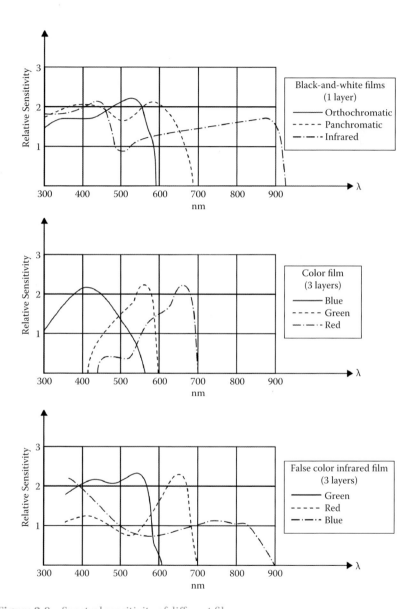

Figure 2.8   Spectral sensitivity of different films.

The colors may be generated artificially by the projection of images of different gray levels in the three primary colors (color additive process). This is the case for a computer or television screen, where three images filtered for the three primary colors are added.

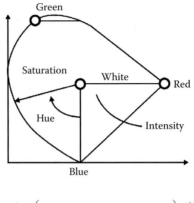

$$\begin{pmatrix} I \\ V_1 \\ V_2 \end{pmatrix} = \begin{pmatrix} 1/3 & 1/3 & 1/3 \\ -1/\sqrt{6} & -1/\sqrt{6} & 2/\sqrt{6} \\ 1/\sqrt{6} & -1/\sqrt{6} & 0 \end{pmatrix} \cdot \begin{pmatrix} d_3 \\ d_2 \\ d_1 \end{pmatrix};$$

$H = tg^{-1}(V_2/V_1);$

$S = \sqrt{V_1^2 + V_2^2};$

with I = Intensity

H = Hue

d_1 = Density in blue

d_2 = Density in green

d_3 = Density in red

Figure 2.9   Chromaticity diagram.

In most photographic work, a color subtractive process is used by absorption filters. This is based on the subtraction of a color from white:

$$\text{White} - \text{Red} = \text{Cyan}$$
$$\text{White} - \text{Green} = \text{Magenta}$$
$$\text{White} - \text{Blue} = \text{Yellow}$$

In color or false color films, three separate film layers are used, which have been sensitized and filtered for the three principal spectral ranges. A color film has a blue sensitive layer followed by a yellow filter, then a green sensitive layer, and finally a red sensitive layer attached to a film base.

Let us now look at the exposure and development process. For color reversal films producing a diapositive, the exposure with blue, green, and red light will initiate the creation of metallic silver in the respective layers. The film is subjected to a black-and-white development. This develops the metallic silver in the respective layers. The result is a black-and-white negative, which is still light sensitive in the unexposed layers. A short subsequent illumination of the film will therefore create silver in the previously unexposed layers. This silver is color developed with color dyes in complementary colors (cyan for red, magenta for green, and yellow for blue). Bleaching of the film will convert all noncolor coupled metallic silver into soluble silver salts. Thus, a film will be the result, which in transparent viewing will yield the original object colors.

Color negative films possess the same three layers, which expose metallic silver in the respective layers. However, the generated silver is directly developed with the color dyes in complementary colors, creating the oxidization product, which cannot be removed by the bleaching process. Thus, a color negative will result, which when copied by the same process can yield a color image in the original colors.

False color photography utilizes three layers sensitive to green, red, and near infrared. The development process is similar to that of color reversal film. The green sensitive layer appears blue in the resulting diapositive, the red sensitive layer in blue, and the infrared sensitive layer in red. The false color film is particularly useful in interpreting the health of vegetation.

Digital Imaging

The irradiated element of a digital imaging system is a photodiode. This charge-coupled device (CCD) consists of a metallic electrode on a silicon semiconductor separated by an oxide insulator. An incoming photon is attracted by the positive voltage of the electrode. A charge is thereby created at the semiconductor acting as a capacitor. The charge is proportional to the number of photons arriving at the electrode (see Figure 2.10).

CCDs are now available at a pixel size up to 5 µm. They may be grouped into linear arrays.

To read out the charges of the diodes of a linear array, a voltage is applied, so that the charge of one diode is shifted to the next one along. This process is repeated until the charges of the whole line are shifted to a storage array as an analogue video signal (see Figure 2.11).

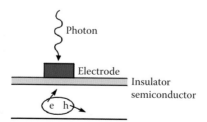

Figure 2.10   CCD detector.

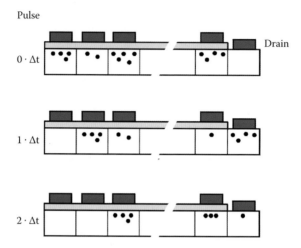

Figure 2.11   Charge transfer in a CCD array.

This video signal can be captured using a tuning signal up to 30 times per second by an analogue to digital converter, called a "frame grabber." The frame grabber captures the exposure as a digital gray level value for later digital processing.

CCD area sensors are composed of several linear arrays arranged in a matrix fashion. The readout into a storage zone occurs line by line in parallel. While the signal is digitally converted in the frame grabber, a new image can be exposed.

The resolution of CCD sensors is limited by the sensor area and by the distance between the sensor elements. These determine a maximal sampling frequency. The geometrical properties of the sensor are governed by the precision with which the sensor elements may be positioned in the image plane.

In contrast to photography, which has a logarithmic D–log H curve, digital sensors have a linear D–H response. As a rule, the digital quantization of the gray level signals is in 2^8 or 256 bits. Newer sensors permit quantization to 2^{11} or 2048 gray levels.

IMAGING SENSORS

Aerial Survey Cameras

Aerial survey cameras must meet stringent requirements to reach high resolution and low geometric distortion for the image. This is achieved through the design of a suitable objective, as shown in Figures 2.12 and 2.13.

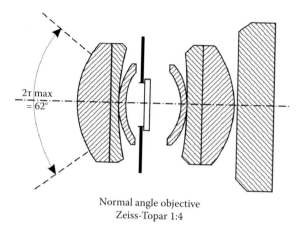

Normal angle objective
Zeiss-Topar 1:4

Figure 2.12   Normal angle objective.

The objective is placed into a vibration dampened mount in a vertical hole in the floor of an aircraft. The camera has a rigid frame in the image plane carrying the fiducial marks, which permit the location of the principal point of the camera. The knowledge of the principal point in the image plane is important, as it defines the interior orientation for the application of the perspective laws according to which an image can be geometrically restituted. The relation of an objective with respect to the image plane is shown in Figure 2.14.

The aerial film of up to 120 m length and 24 cm width is contained in a cassette mounted on top of the image plane. It contains a pressure plate, which at the time of the exposure is pressed against the image frame. At that time, the film guided through the image plane is pressed onto the pressure plate by a vacuum system to ensure the flatness of the exposed film. For that purpose,

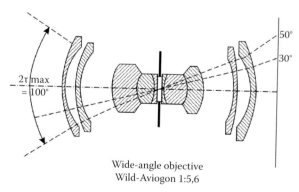

Wide-angle objective
Wild-Aviogon 1:5,6

Figure 2.13   Wide-angle objective.

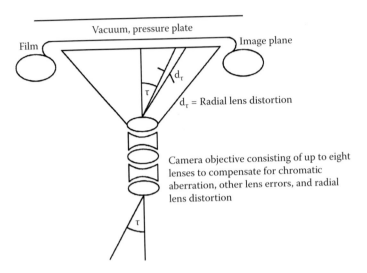

Figure 2.14   Image plane and objective.

the pressure plate is equipped with suction holes. After the exposure, the film pressure is released to allow forward motion of the film for the next exposure by about 25 cm, accounting for the standardized image format of 23 × 23 cm.

Photography is controlled by a rotating shutter, permitting a simultaneous exposure of all parts of the image every 2 seconds for a duration between 1/100 and 1/1000 of a second. The more recent camera types (Leica RC 30 and Zeiss RMK TOP) permit a forward motion of the film during the exposure to enable the exposure of the same terrain for longer intervals, while the airplane moves forward, permitting the use of high-resolution film, which, due to its smaller silver halide grains, needs a longer exposure time for an exposure falling into the linear part of the D–log H curve. This image motion compensation generally improves the achievable resolution (see Figure 2.15).

Figures 2.16 and 2.17 show the aerial camera types produced by the manufacturers LH-Systems and Z/I Imaging.

A systematic survey of the terrain is only possible if the camera frame is oriented in parallel in the flight direction. To eliminate the crab angle between camera orientation and the flight axis, the camera is rotated within the mount by a mechanism (see Figure 2.18).

Older cameras control this movement by the flight photographer using an overlap regulator consisting of a ground glass image plane onto which the ground is continuously imaged while the aircraft moves forward. The overlap regulator permits the control of exposure interval of more than 2 seconds, so that a particular overlap of ordinarily 60% in the flight direction is achieved. Newer cameras

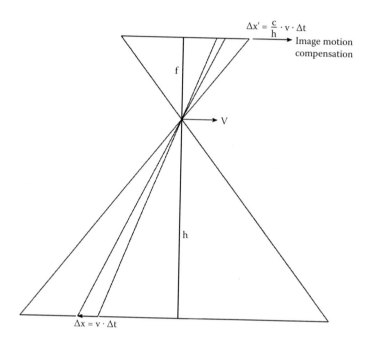

$$\Delta x' = \frac{c}{h} \cdot v \cdot \Delta t$$

Image motion compensation

f

V

h

$\Delta x = v \cdot \Delta t$

Figure 2.15   Image motion compensation.

Figure 2.16   The LH Systems RC30 camera.

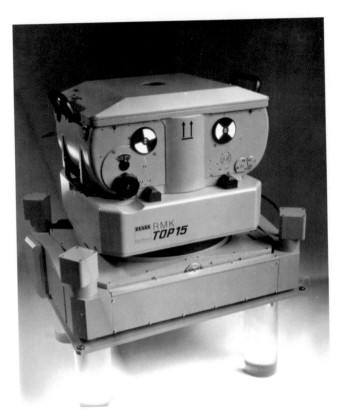

Figure 2.17 The Z/I RMK TOP camera. (Courtesy of Z/I Imaging Corporation, Oberkochen, Germany.)

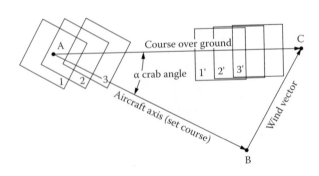

Figure 2.18 Crab angle.

TABLE 2.3   ANALOGUE CAMERA SYSTEMS

Manufacturer	Name of Camera	Type of Objective	Image Angle, $2\tau_{max}$	Principal Distance, f	Ratio d/f	Maximal Image Distortion
LH Systems	RC 30	UAG-S	90°	153 mm	1:4	4 μm
LH Systems	RC 30	NATS-S	55°	300 mm	1:5.6	4 μm
Z/I Imaging	RMK-TOP 15	Pleogon A3	93°	153 mm	1:4	3 μm
Z/I Imaging	RMK-TOP 30	TOPAR A3	54°	305 mm	1:5.6	3 μm

control orientation of the camera mount by the crab angle and the set overlap percentage by an automatic sensor, which checks the aircraft motion over the ground h/v. The time interval between exposures, t, for a given overlap, q, is calculated by

$$t = \frac{s}{v} = \frac{h}{v} \cdot \frac{a'}{f}\left(1 - \frac{q}{100}\right)$$

in which s is the ground distance between exposures, v the velocity of the aircraft, a' the image size (23 cm) and f the chosen focal length (equal to the principal distance for imaging at infinity).

The major aerial survey camera types in use are contained in Table 2.3, along with their characteristics.

Photogrammetric camera developments had their main roots in Central Europe starting during World War I at Carl Zeiss in Jena, Germany. Aerial cameras were later also produced in Italy, France, Britain, and the United States. The other main producer besides Zeiss became Wild Heerbrugg in Switzerland, which became a major aerial camera producer between the 1930s and the 1970s. Wild became Leica Geosystems in the 1980s. Camera development in Germany after World War II was continued in West Germany by Carl Zeiss Oberkochen since the 1950s and in East Germany since the 1960s by Carl Zeiss Jena.

Now the efforts of Leica and the reunited Zeiss of the 1990s have all been consolidated by the Swedish Investment Company Hexagon after the millennium.

A major competitor has become Vexcel in Austria under the ownership of Microsoft. But also new progressive companies have emerged, such as Visionmap in Israel, IGI in Germany, and Rollei (Trimble) in Germany.

While camera development in the past has been single-sensor oriented, the current tendency is toward integration of sensor systems and a future collaboration of sensors. Oblique imagery, introduced by Pictometry and Multivision as stand-alone systems, may now be provided by inexpensive additional

Airborne sensors

Leica ADS Pushbroom	Z/I DMC II Frame	Leica RCD Frame	Leica ALS
Imaging	Imaging	Imaging	LIDAR

Figure 2.19 Some manufactured Hexagon cameras. (From Leica Geosystems, Heerbrugg, Switzerland, for Hexagon.)

sensors with a smaller number of pixels than those required for a class A type high-resolution aerial mapping camera.

Sensors of various resolutions may now be accommodated to different platforms:

- Land based (portable on a van or fixed)
- Unmanned aerial vehicle (UAV) platforms
- Airborne
- Space

All types may be assisted by navigational sensors, such as GPS (GNSS) and an inertial measuring unit (IMU).

A few illustrations of the latest Hexagon sensors are added here (Figure 2.19):

1. Z/I (Zeiss/Intergraph) DMC frame camera
2. Leica RDC frame camera
3. Leica ADS Pushbroom scanner
4. Leica ALS Lidar

All of them may be mounted in an aircraft (Figure 2.20).

The characteristics of the Z/I Imaging DMC are shown in Figure 2.21. The DMC comes in three versions:

- DMC 140 with 4 × 42 Mpixel MS CCDs of 7.2 μm size and 1 × 140 Mpixel pan CCD of 7.2 μm size giving an image of 12096 × 11200 pixels (10 cm GSD at h = 400 m)

Figure 2.20 Installed Hexagon cameras. (From Leica Geosystems, Heerbrugg, Switzerland, for Hexagon.)

- DMC 230 with 4 × 42 Mpixel MS CCDs of 7.2 μm size and 1 × 230 Mpixel pan CCD of 5.6 μm size giving an image of 15552 × 14144 pixels (10 cm GSD at h = 1600 m)
- DMC 250 with 4 × 42 Mpixel MS CCDs of 7.2 μm size and 1 × 250 Mpixel pan CCD of 5.6 μm size giving an image of 16768 × 14016 pixels (15 cm GSD at h = 3000 m)

DMC II

A Product of Z/I Imaging

- Multi spectral sensor, RGB and IR
- 2:1 pan-sharpened color resolution
- FMC forward motion compensation
- 2 second frame rate
- B/H ratio of 0.36 @ 60% overlap
- 4x42 MPixel, 7.2 um MS CCD
- 1x140 MPixel, 7.2um PAN CCD
- Finished Image Size: 12,096 x 11,200
- 10cm GSD @ ~400 m flying height

Figure 2.21 Z/I Imaging DMC2. (From Leica Geosystems, Heerbrugg, Switzerland, for Hexagon.)

DMC II

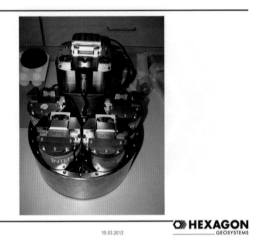

Leica ZI | 15.03.2013 | ⦿ HEXAGON GEOSYSTEMS

Figure 2.22   Mounting of 5 DMC2 camera parts in frame. (From Leica Geosystems, Heerbrugg, Switzerland, for Hexagon.)

The five DMC camera parts are mounted in one frame (see Figure 2.22).

The Leica RCD 30 is a coregistered RGB and IR multispectral camera, either with a 50 or 80 mm principal distance. The camera has 2 × 60 Mpixel CCDs of 6 μm size for RGB and NIR with an image size of 8956 × 6708 pixels (15 cm GSD at h = 1000 m).

The Leica Pushbroom Scanner ADS 80 scans perpendicular to the flight line with 24000 pixels in three simultaneous scans:

- Forward in panchromatic
- Nadir in panchromatic and multispectral
- Aft in multispectral or panchromatic (see Figure 2.23)

The Leica ALS 70 laser scanner has a scan rate of 500 KHz for use at a flying height up to 3500 m or 250 KHz up to h = 5000 m. Scan patterns may be selected for sine, triangle, or raster scans. The field of view is adjustable up to 75° (full angle). The laser return image operated from aircraft is shown in Figure 2.24.

The scanner can also be operated from a mobile van (Figure 2.25), with a result shown in Figure 2.26.

Microsoft-Vexcel, Austria, has developed the UltraCam camera. It came on the market as UltraCam D, which subsequently was improved to UltraCam X and UltraCam Xp and finally UltraCam Eagle with 20010 pixels across (see Figure 2.27). The UltraCam Eagle has an exchangeable lens system with principal distances of either 80 mm or 210 mm (Figure 2.28).

Pushbroom 24k HiRes mode

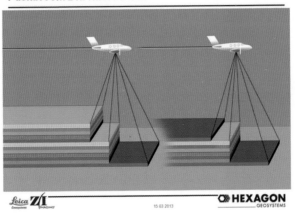

Figure 2.23   Leica Geosystems ADS80 scan. (From Leica Geosystems, Heerbrugg, Switzerland, for Hexagon.)

Processing of UltraCam images is carried out by the UltraMap software to various levels. It starts with the download of the images from the camera into mass storage immediately after the flight, with subsequent offline processing to higher levels, including aerial triangulation, by an efficient Microsoft processing system "Dragon Fly" (Figure 2.29).

It permits the use of multiray photogrammetry, based on 80% longitudinal and 60% lateral overlap, by which up to 12 rays are created for each ground pixel. This helps to avoid occlusions (see Figures 2.30 and 2.31).

ALS70

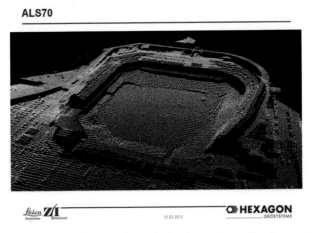

Figure 2.24   Laser image of ALS70. (From Leica Geosystems, Heerbrugg, Switzerland, for Hexagon.)

Figure 2.25   Mobile van operation of ALS70. (From Leica Geosystems, Heerbrugg, Switzerland, for Hexagon.)

This gives the possibility to generate "true orthophotos," or as Microsoft prefers to call them, "DSM orthophotos" (see Figure 2.32).

Figure 2.33 is an example of a 10 cm GSD MS image, and Figure 2.34 shows the DSM derived for that image by multiray matching. Figure 2.35 shows such a color-coded DSM for the cathedral area of the City of Graz, Austria.

Figure 2.26   Laser image from mobile van. (From Leica Geosystems, Heerbrugg, Switzerland, for Hexagon.)

Figure 2.27   Vexcel UltraCam cameras. (From Microsoft-Vexcel, Graz, Austria.)

Image Quality of Aerial Survey Cameras

Image quality in photographic images is determined by the contrast between two image objects with the irradiances E_0 and E_N or its reflection coefficients ρ_o and ρ_N. The object contrast is therefore:

$$K = \frac{E_N}{E_0} = \frac{\rho_N}{\rho_o}$$

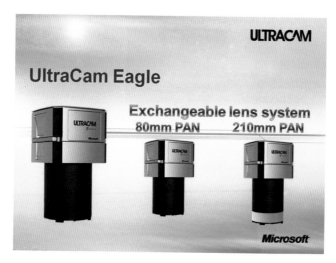

Figure 2.28   UltraCam Eagle. (From Microsoft-Vexcel, Graz, Austria.)

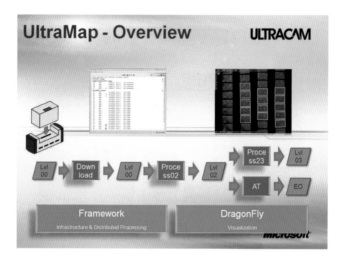

Figure 2.29   UltraMap Processing System. (From Microsoft-Vexcel, Graz, Austria.)

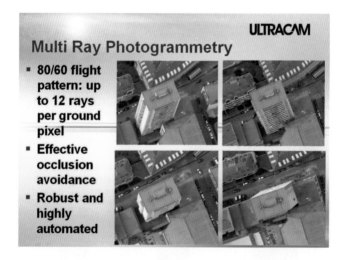

Figure 2.30   Multi Ray photogrammetry. (From Microsoft-Vexcel, Graz, Austria.)

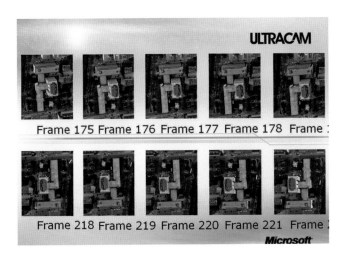

Figure 2.31   Overlapping frames. (From Microsoft-Vexcel, Graz, Austria.)

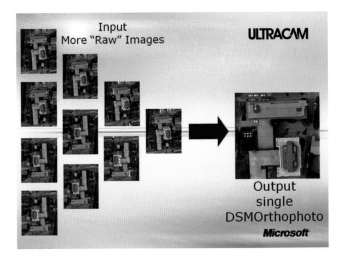

Figure 2.32   DSM orthophotos. (From Microsoft-Vexcel, Graz, Austria.)

Figure 2.33　10 cm GSD UltraCam image. (From Microsoft-Vexcel, Graz, Austria.)

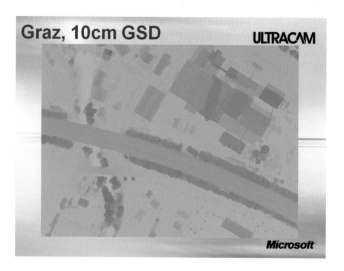

Figure 2.34　DSM derived. (From Microsoft-Vexcel, Graz, Austria.)

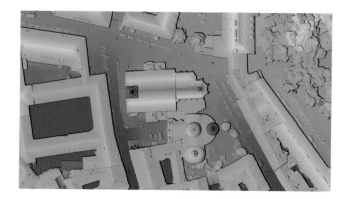

Figure 2.35 Color-coded DSM of Graz, Austria. (From Microsoft-Vexcel, Graz, Austria.)

Modulation is defined by:

$$M = \frac{\rho_N - \rho_0}{\rho_N + \rho_0} = \frac{K-1}{K+1}$$

This object contrast or the corresponding object modulation is visible in the image as image contrast K' or as image modulation M' depending on the transparencies τ_o and τ_N of the imaged object contrast

$$K' = \frac{\tau_N}{\tau_o}$$

and

$$M' = \frac{\tau_N - \tau_o}{\tau_N + \tau_o} = \frac{K'-1}{K'+1} = \frac{10^{\Delta D} - 1}{10^{\Delta D} + 1}$$

with

$$\Delta D = \log \frac{\tau_N}{\tau_o}$$

the contrast density difference and modulation can therefore be converted into corresponding values shown in Table 2.4.

TABLE 2.4   CONTRAST, DENSITY DIFFERENCE, AND
MODULATION

Contrast	1000:1	2:1	1.6:1
Density difference	3.0	0.3	0.2
Modulation	0.999	0.33	0.23

The contrast transfer, C', and the modulation transfer, C, through the imaging system is therefore:

$$C' = \frac{K'}{K} \quad \text{and} \quad C = \frac{M'}{M}$$

Both C' and C are functions of the spatial distance, expressed in terms of a spatial frequency v. This is shown in Figure 2.36 for the frequency dependent modulation transfer function.

Image quality in a combined optical–photographic system is deteriorated by the imaging process if the modulation transfer function, C, is <1. Only for an extremely high contrast (1000:1) does C become nearly 1.

Image quality in a photographic system, other than by contrast, is influenced by the following components:

- Optical system
- Film
- Image motion during exposure

Each component is characterized by its own modulation transfer function, C_{optics}, C_{film}, $C_{\text{image motion}}$. The total effect is C_{total}:

$$C_{\text{total}} = C_{\text{optics}} \cdot C_{\text{film}} \cdot C_{\text{image motion}}$$

The typical components for an aerial camera system are shown in Figure 2.37.

The resolution of an optical system is first limited by diffraction. The resolution A in line pairs per millimeter (lp/mm) can be expressed by

$$A = \frac{1000 \cdot d}{24 \cdot \lambda \cdot f}$$

in which d is the diameter of the diaphragm, λ the wavelength, and f the focal length. For a d/f value of 4, this amounts to 120 lp/mm. Other limiting effects stem from the composition of the lens system causing lower actual resolution.

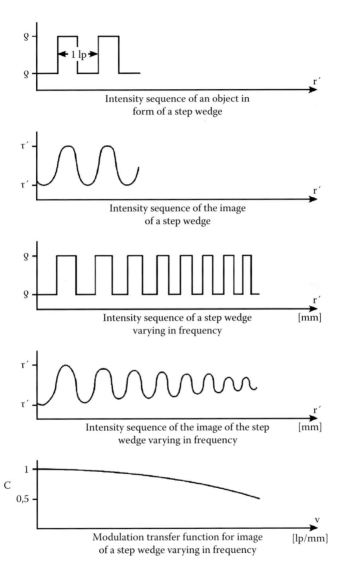

Intensity sequence of an object in
form of a step wedge

Intensity sequence of the image
of a step wedge

Intensity sequence of a step wedge   [mm]
varying in frequency

Intensity sequence of the image of the step   [mm]
wedge varying in frequency

Modulation transfer function for image   [lp/mm]
of a step wedge varying in frequency

Figure 2.36   Modulation transfer function.

The photographic film resolution depends on its grain size. High sensitive film contains coarse silver halide particles with a resulting low resolution. Low sensitive film has small grain size of the particles resulting in high resolution. The resolution of typical aerial films is shown in Table 2.5.

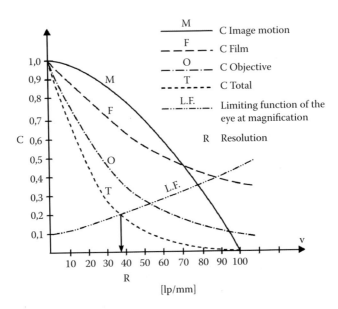

Figure 2.37 Modulation transfer function components for an aerial photographic system.

Image motion $\Delta a'$ in micrometers during exposure depends on the exposure time, Δt, on the platform movement over the ground (v) in kilometers per hour, the focal length (f), and the flight altitude (h):

$$\Delta a' = \frac{f}{h} \cdot \frac{\Delta t}{3600} \cdot v$$

Under the assumption that the modulation assumes a value of 0 at a value of

$$a' = \frac{1000}{\Delta a'}$$

in line pairs per millimeter, the modulation transfer function due to image motion becomes:

$$C_{\text{image motion}} = \cos\left(\frac{v}{2a'}\right)$$

The human eye determines a limiting function for which contrasts can still be recognized at low modulation. Intersecting this limiting function with C_{total} yields the interpretable resolution in line pairs per millimeter.

TABLE 2.5   AERIAL SURVEY FILMS

Manufacturer	Name	Sensitivity	Resolution of Contrast 1000:1	Resolution at Contrast 1.6:1	Type
Agfa	Aviphot pan 200 S PE1	125–200 C ISO (23 DIN)	130 lp/mm	50 lp/mm	Panchromatic
Agfa	Aviphot pan 80	64–100 ISO (20 DIN)	287 lp/mm	101 lp/mm	Panchromatic
Agfa	Aviphot color X100 PE	125–160 ISO (22 DIN)	140 lp/mm	55 lp/mm	Color negative
Agfa	Aviphot color N400 PE	400–640 ISO (28 DIN)	130 lp/mm	35 lp/mm	Color negative
Agfa	Aviphot chrome 200 PE1/PE3	200 C ISO (24 DIN)	110 lp/mm	50 lp/mm	Color reversal
Kodak	Aerographic 2402	160 ISO (23 DIN)	130 lp/mm	55 lp/mm	Panchromatic
Kodak	Aerographic 2403	640 ISO (29 DIN)	100 lp/mm	40 lp/mm	Panchromatic
Kodak	Double X Aerographic 2405	400 ISO (27 DIN)	125 lp/mm	50 lp/mm	Panchromatic
Kodak	Panatomic X 3412	40 ISO (17 DIN)	400 lp/mm	125 lp/mm	Panchromatic
Kodak	Aerocolor 2445	64 ISO (19 DIN)	80 lp/mm	40 lp/mm	Color negative
Kodak	Aerochrome 2448	32 ISO (16 DIN)	80 lp/mm	40 lp/mm	Color reversal
Kodak	Aerocolor SO 846	160 ISO (23 DIN)	100 lp/mm	63 lp/mm	Color negative

The atmosphere generally diminishes the object contrast for all objects sensed by aerial imaging. If two objects consisting of sand with a reflection coefficient ρ_{sand} = 30 and coniferous forest with a reflection coefficient ρ_{forest} = 1 are observed, the object contrast, K, may be expressed as:

$$\frac{\rho_{sand}}{\rho_{forest}} = \frac{30}{1} = 30 = K$$

The image contrast is then affected by the diffused atmospheric light of 3%. Thus:

$$K' = \frac{30+3}{1+3} = \frac{33}{4} = 8 = K'$$

This illustrates that resolutions of generally low-contrast aerial photographs should not be compared at a contrast of 1000:1, but at a contrast of 1.6:1, for which the modulation of 1 diminishes to 0.23. Thus, the modulation transfer function, as shown in Figure 2.38, is flattened, and its intersection with the limiting function of the eye is lowered in resolution.

Aerial survey camera systems with image motion compensation are, in practice, achieving a resolution of between 40 to 50 lp/mm for black-and-white images and of between 30 and 40 lp/mm for color images.

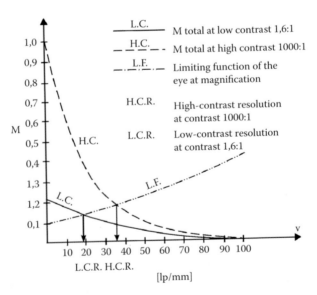

Figure 2.38   Resolution at low contrast.

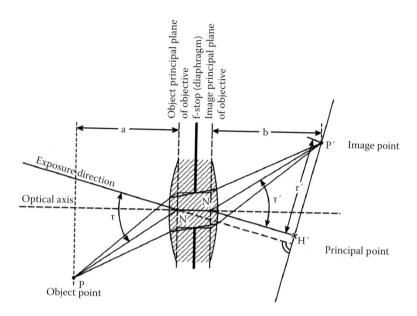

Figure 2.39   Imaging through an objective.

Another property of aerial survey cameras is the ability to generate photographic images with minimum geometric distortion. A primary source of geometric distortion stems from the objective (Figure 2.39).

Radial distortion, $\Delta\tau$, is the angular difference between the angle of exposure direction and object direction, τ_i, and their imaged difference τ_i':

$$\Delta\tau_i = \tau_i - \tau_i'$$

Distortion can also be expressed as a radial distance difference in the image plane

$$\Delta_r' = r_i - r_i'$$

with r_i' measured from the principal point, determinable from the fiducial marks of the camera and $r_i = f \cdot tg\ \tau_I$ (see Figures 2.40 and 2.41). N marks the entrance of exit modes of rays along the optical axis through the objective.

Camera manufacturers have tried to minimize the distortion in a factory calibration procedure in which the angles τ_i' are measured through the objective by a goniometer. The fiducial marks of the camera, determining the principal point, are fixed so that a minimum radial and tangential distortion results.

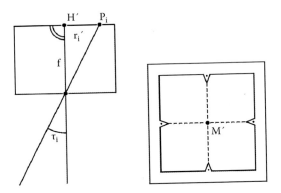

Figure 2.40 Principal point.

Camera manufacturers issue calibration certificates in which the attained radial distortions are listed. They are in the order of 2 µm and usually never exceed 4 µm in the image plane. Tangential lens distortions are generally only one-third of the radial distortions and thus negligible for factory calibrations (see Figure 2.42).

A second source of distortion is the film. If the pressure plate vacuum has properly worked, the image plane can be considered as flat. Otherwise distortions of

$$\Delta r_i' = \frac{r'}{f} \Delta f = \Delta f \cdot tg\tau'$$

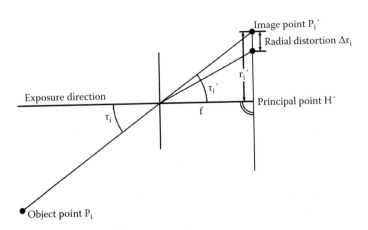

Figure 2.41 Definition of radial distortion.

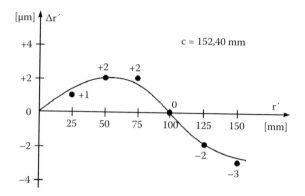

Figure 2.42   Radial distortion of a photogrammetric objective.

will result. Δf is the deviation of the image point on the film from the ideal focal plane at the principal distance f.

More serious are the geometric distortions of the film caused by the wet developing and drying process. For this reason, each camera contains between four to eight fiducial marks. If their images are measured, a correction of the film deformations in the direction of the film and perpendicular to it can be made during the restitution. Fortunately, these types of film deformations remain constant for the entire film, so that they can be accounted for by parameters in the restitution process.

Optomechanical Scanners

An optomechanical scanner contains a single sensor element, which permits the recording of the irradiance of a ground pixel. A rotating mirror scans the terrain, so that a whole line of ground pixels can be recorded in a time sequence. The next scan of a forward-moving platform records the adjacent line of ground pixels. Thus, the scanning mechanism during forward motion permits a recording of a whole image. The scanning principle is shown in Figure 2.43.

For an altitude, h, a field of view of the scanned pixel, ω, the total angular scan width, Ω, and the scan angle from the vertical, α, the ground pixel size, a, in the scan direction becomes:

$$a = \frac{\omega \cdot h}{\cos^2 \alpha}$$

The ground pixel size, b, in the line of flight, b, is likewise:

$$b = \frac{\omega \cdot h}{\cos^2 \alpha}$$

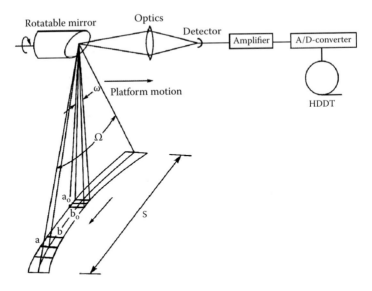

Figure 2.43   Operation of an optomechanical scanner.

The swath of the scan, s, becomes:

$$s = 2h \cdot tg \frac{\Omega}{2}$$

The scan frequency, v, is a function of the platform velocity, v_g:

$$v = \frac{v_g}{\omega \cdot h}$$

Optomechanical scanners have been utilized for airborne and for satellite multispectral scanners, as shown in Figure 2.44.

Instead of a single photodiode, a linear array was used as a sensor. The spectral separation of the incoming energy was achieved by the diffraction of a prism in the optical path, so that different bands of wavelengths could be recorded at the array at a particular time.

The scanner even permitted the recording of thermal energy with the use of a dichroitic separation of the ray. Thermal energy could then be collected on a thermally cooled (77 or 5 K) far-infrared sensitive mercury-doped CD-telluride or germanium detector.

The characteristics of such detectors are shown in Table 2.6.

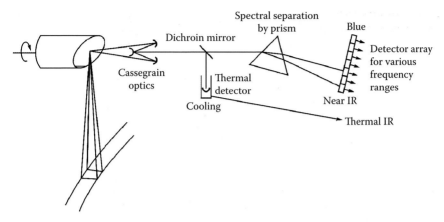

Figure 2.44   Operation of a multispectral scanner.

Laser Scanners

It is possible to direct laser light impulses to the terrain along the principles of electromechanical scanning. Part of the reflected radiation returns to the laser scanner. This gives the possibility for measuring the time between the emission of the pulse and the first and the last return, as well as the energy received.

This permits the use of the device as a laser altimeter for any spot reached by the laser pulse. If the position of the sensor is measured by in-flight GPS, and its orientation by inertial navigation devices, then it becomes possible to determine the three-dimensional position of the reflection point.

On this principle, laser scanners, such as the system operated by Toposys, have been built and operated: it is a pulsed fiber scanner, operated at an airborne altitude of less than 1600 m with a laser wavelength of 1.55 μm. It emits pulses every 5 nsec. The scan frequency is 650 Hz and the pulse repetition rate is 83000 Hz. This permits, within a field of view of $7°$ from the vertical, the measurement of laser reflection points of a density of three points per square meter on the surface.

TABLE 2.6 OPTOMECHANICAL DETECTORS					
Scanner	**Platform**	**ω**	**Ω**	**Number of Visual and Near IR Channels**	**Number of Thermal Channels**
Daedalus DS 1200	Aircraft	2.5 mrad	77°	2	1
Landsat MSS	Satellite	0.087 mrad	11.6°	4	0
Landsat TM	Satellite	0.024 mrad	11.6°	6	1

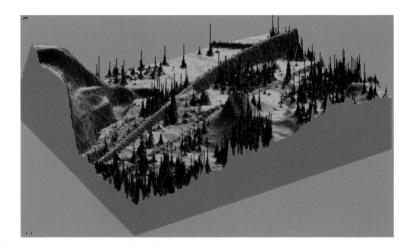

Figure 2.45   Original laser scan. (Courtesy of the Institute for Photogrammetry and GeoInformation, University of Hannover, Germany.)

The recording of the first and last pulse received from the point permits the judging of the thickness of the vegetation cover of the terrain. The swath covered in a flight strip is then up to 390 m, and elevation measurements are possible within a relative accuracy of 2 cm and an absolute accuracy of 15 cm for a digital elevation model. Leica-Helava Systems produces a laser scanner ALS 40 for the generation of digital elevation models. Figures 2.45 and 2.46 show the laser signals received in original and filtered forms.

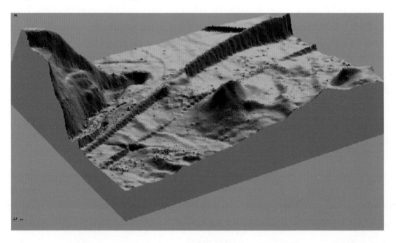

Figure 2.46   Filtered laser scan. (Courtesy of the Institute for Photogrammetry and GeoInformation, University of Hannover, Germany.)

It is possible to use the intensity of the received signal to construct an image with respect to the digital elevation model, even though this use of the laser scanner is still experimental (see www.toposys.com and www. geolas.com).

Optoelectronic Scanners

Optoelectronic imaging is possible when an image created through an optical system is created on a linear or a matrix array of CCD sensors. Linear arrays of sufficient length are easier to assemble than matrix arrays. Therefore, the electrooptical scanner, as shown in Figure 2.47, has frequently been applied in satellite sensors using the push-broom principle.

A push-broom scanner has the linear array oriented perpendicular to the platform motion. A detector element of the dimension a' perpendicular to the flight direction will cover a ground pixel dimension, a, according to

$$a = \frac{h}{f} \cdot a'$$

with h being the platform altitude and f the focal length of the optics. In-flight direction, the detector element of the dimension b' in this direction, is likewise imaged at a ground pixel dimension, b:

$$b = \frac{f}{h} \cdot b'$$

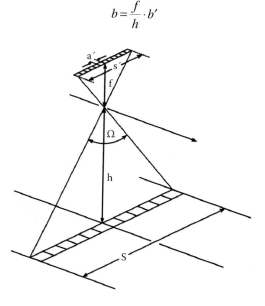

Figure 2.47   Optoelectronic scanner.

Name	ω	Ω	Number of Channels	Type
SPOT-P	0.012 mrad	4.2°	1	Panchromatic
SPOT.XS	0.024 mrad	4.2°	3	Multispectral

TABLE 2.7   PARAMETERS OF THE SPOT ELECTROOPTICAL SCANNER

To assemble an image, the exposure, Δt, must be chosen proportional to the velocity of the platform with respect to the ground, v_g:

$$b = v_g \cdot \Delta t$$

The swath of the push-broom sensor, s, becomes

$$s = \frac{h}{f} \cdot s' = 2h \cdot tg \frac{\Omega}{2}$$

with s' being the length of the linear array pointing symmetrically to the vertical. An example is the sensor for the French Spot satellite, characterized in Table 2.7.

The Spot satellite sensor not only uses a panchromatic array with 10 m ground pixels but, in parallel to that, three farther arrays of half resolution, yielding 20 m ground pixels for the filtered spectral bands of green, red, and near infrared.

One Spot satellite sensor may be inclined sideways in a programmed mode in steps from –27° to +27° to cover any point on earth, subject to cloud cover permitting it, in 5 days. Three operating Spot satellites can do so in 1 day.

Digital electrooptical cameras have also been constructed by the DLR for use on Mars. After a Russian spacecraft failed on launch, the DLR constructed an aircraft version of that camera. Its design has been manufactured by the Leica-Helava Systems Company for aerial surveys as the ADS 40 airborne digital sensor (www.lhsystems.com) (see Figure 2.48). A raw and a rectified image of that sensor are shown in Figures 2.49 and 2.50.

The digital sensor of the Zeiss-Intergraph Imaging Company, the digital modular camera (DMC), has split objectives imaging onto matrix sensors of seven image planes (Figures 2.51 and 2.52). The images are resampled and reconstructed in a single frame by software (www.ziimaging.com).

The realization of a stereo electrooptical scanner is due to O. Hofmann at MBB. It was built for use in the space shuttle and MIR for the MOMS 02 and MOMS 02-P satellite missions (Figure 2.53).

The MOMS 02 sensor possesses a vertical CCD line with 5 to 6 m ground pixel and two forward- and aft-looking CCD arrays with 15 to 18 m pixels in the same image plane in the panchromatic range. Parallel to the vertically looking panchromatic CCD, there are three additional filtered multispectral arrays at 15 to 18 m ground pixels.

Figure 2.48 The LH Systems ADS 40 digital camera. (Courtesy of LH Systems [Leica Geosystems], San Diego, California. © Leica Geosystems, 2000.)

For stereo sensing, only the vertical panchromatic and the two forward- and aft-looking CCD lines are used. Further developments of this principle have been introduced in the Spot 5 satellite of 2002 yielding a ground pixel of 2.5 m, paired with an equivalent array oriented at a fixed angle.

Image Spectrometers

It is possible to design an optoelectronic scanner with an optoelectronic array of photosensitive elements in conjunction with diffractive gratings in such a manner that, in combination with a continuous variable optical filter, narrow bands of only 10 mm wavelength are projected onto the array resulting in continuous spectral signatures in the visible and infrared spectrum. These so-called hyperspectral devices have the ability to image in up to 224 spectral channels, which can be compared with object libraries. The spatial resolution of these hyperspectral devices has, of course, to be reduced in accordance with the reflected energy available. For the AVNIR, which operates in 60 visible and near-infrared bands at wavelengths between 430 and 1012 nm in 10 nm increments, a ground sample distance of 0.8 m can be reached from an altitude of 1600 m. The AVIRIS spectrometer of NASA-JPL is operated from high-altitude aircraft ($h = 20$ km). It has 224 spectral channels at intervals of 10 nm between 400 nm and 2450 nm wavelength. Its ground pixel size at that altitude is 17 m and the swath 11 km (http://aviris.jpl.nasa.gov/).

Figure 2.49 Raw ADS 40 image. (Courtesy of LH Systems [Leica Geosystems], San Diego, California. © Leica Geosystems, 2000.)

NASA has launched a TRW-built 200 channel image spectrometer Hyperion on the EO-1 spacecraft in the year 2000, with 30 m ground pixels at a swath of 7.5 km.

Oblique Imaging

The availability of lower-cost digital camera technology has generated the interest to design new multiple camers systems, by which not only vertical imagery is acquired, but also oblique images from front and aft and left and right camera orientations.

Oblivision and Multivision designed in Israel; and Pictometry, an oblique system in the United States, are such examples. Pictometry was widely used for Homeland Security activities.

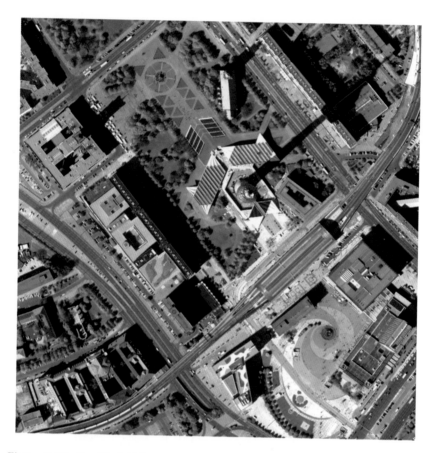

Figure 2.50 Rectified ADS 40 image. (Courtesy of LH Systems [Leica Geosystems], San Diego, California. © Leica Geosystems, 2000.)

More recently, camera manufacturers, such as Leica Geosystems, Microsoft Vexcel, and ISI in Germany have also put such multicamera systems on the market. These can effectively be used not only for the creation of digital city models, but the oblique images also permit the viewing of image details of the façades of houses.

Radar Imaging

The natural radiation in the microwave range of the electromagnetic spectrum is generally too weak to be useful for imaging. Therefore, passive sensing is rare. Radar imaging, therefore, utilizes an active sensor, generating the transmitted and reflected energy in the microwave region.

Figure 2.51 The Z/I DMC digital camera. (Courtesy of Z/I Imaging Corporation, Huntsville, Alabama.)

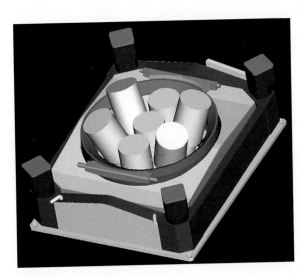

Figure 2.52 The Z/I DMC split objectives. (Courtesy of Z/I Imaging Corporation, Huntsville, Alabama.)

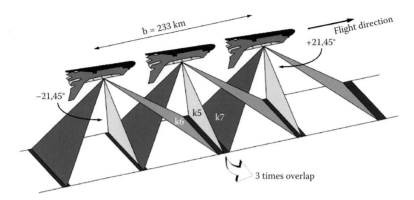

Figure 2.53   MOMS stereo scanner.

Radar systems have principally been built in three wavelength regions:

- X band, λ = 2.4 to 3.8 cm (8000 to 12500 MHz)
- C band, λ = 3.8 to 7.5 cm (4000 to 8000 MHz)
- L band, λ = 15 to 30 cm (1000 to 2000 MHz)

The X band and the C band in particular have the advantage of cloud penetration. They can therefore operate day and night in an all-weather system.

Let us now look at the simple operating principle of a radar. A transmitter generates a radar pulse composed of a wave signal in the respective band. The duration of the radar pulse is Δt. It is transmitted through an antenna with special propagation characteristics, so that the energy is concentrated in a narrow beam perpendicular to the platform motion. It reaches the ground with the speed of propagation of electromagnetic waves, v,

$$v = \frac{c}{n}$$

with c being the velocity in vacuum and n the propagation coefficient.

At the terrain, the energy is directionally reflected, scattered, or absorbed. The reflected energy retroreflected into the direction of the transmitting antenna is characterized by the radar equation for the radiant flux received

$$\phi = \frac{\phi_o \cdot G^2 \cdot \lambda^2 \cdot \rho_o \cdot A}{(4\pi)^3 \cdot r^4} \cdot 10^{-0.2 \cdot \alpha \cdot r}$$

with the following quantities:

ϕ_o = transmitted radial flux from the antenna in W
G = antenna gain along the direction of transmission
ρ_o = retroreflection or backscattering coefficient of the terrain point

A = reflecting surface in square meters
r_i = distance between antenna and object i
α = coefficient of atmospheric attenuation, which is wavelength λ dependent

The transmitted and backscattered radar pulse of different terrain points along the plane of transmission reaches the antenna at different times, T_i:

$$T_i = \frac{2r_i}{v}$$

The achievable ground resolution of a radar system in the direction perpendicular to the platform motion, a, depends on the duration of the pulse, Δt, and the depression angle, β, between the horizon and the transmitted and reflected ray:

$$a = \frac{v \cdot \Delta t}{2 \cdot \cos\beta}$$

For the reception of the backscattered energy, the antenna is switched by a duplexer from transmission to reception. This permits the recording of the incoming signals as a function of T_i. When the reception from all the terrain points in the plane is completed, the antenna is again switched to the transmitter and a new pulse is sent while the platform has moved forward.

The resolution in the direction of forward motion, b, like for scanners, depends on the platform velocity, v_g, and the time interval, ΔT, between successive pulse transmissions:

$$b = v_g \cdot \Delta T$$

However, since it is difficult to bundle the transmitted energy in one plane, the time interval, ΔT, at which two successive pulses may be transmitted, equally depends on the antenna characteristics. The azimuthal dimension $\Theta°$ of a radar beam in the transmission plane depends on the length, L, in meters of the transmitting antenna, for which the following relation is valid:

$$\Theta° = 60 \cdot \frac{\lambda}{L} \quad \text{and} \quad b = r \cdot \Theta° = \frac{h}{\sin\beta} \cdot \Theta°$$

This limits the azimuthal resolution of a side-looking airborne radar (SLAR), since the length of an antenna depends on the length of a platform (e.g., an airplane). Radar imaging is illustrated in Figure 2.54.

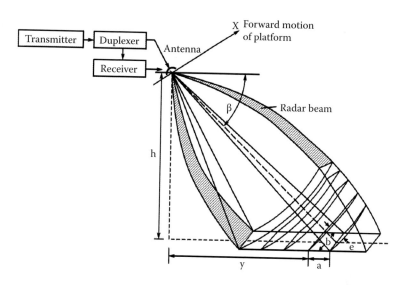

Figure 2.54 Radar imaging.

There is a better possibility to improve the azimuthal resolution with small antennas, which radiate the energy in a wide beam. Since the radar pulse emits a wave at a known frequency, the coherent energy from the reflected target not only permits the uses to determine its intensity but also to use its frequency information. A target's position along the flight determines the Doppler frequency of its backscatter. Targets ahead of the aircraft produce a positive Doppler offset; those behind the aircraft produce a negative Doppler frequency.

Thus, the signal may be geometrically focused to a Doppler frequency of zero. This is analogous to a holographic reconstruction of the wave signals to form an image. The image coordinate in azimuthal direction can be generated as a slant range distance, y'_s,

$$y'_s = \frac{v \cdot T}{2} \cdot m_y = m_y \cdot r_s$$

in which v is the velocity of wave propagation, T is the time interval between emission and reception, and m_y is a scale factor. r_s is the slant range.

The slant range distance, y'_s, can be reduced to a ground range distance, y'_G, for a specified platform elevation, h:

$$y'_G = m_y \cdot \sqrt{r_s^2 - h^2}$$

The image coordinate in flight direction is:

$$x' = m_x \cdot x$$

The scale factor, m_x, is a function of the platform velocity, v_g.

Height differences of the terrain, Δh, cause slant range differences, Δr, or their horizontal projection, Δy. According to Figure 2.55:

$$\Delta r = \sqrt{y^2 + (h - \Delta h)^2}$$

or

$$\Delta y = \sqrt{y^2 - 2\Delta h h + \Delta h^2}$$

Polarization of a Radar Beam

Most side-looking airborne radar (SLAR) or synthetic aperture radar (SAR) systems have antennas emitting the radar pulses in a polarization plane. Most frequently, the horizontal polarization, H, is chosen, but it is also possible to use a vertical polarization, V.

Due to the fact that radar backscattering objects may depolarize the reflected signal in all directions, it becomes possible to receive the radar backscatter in either the horizontal or the vertical polarization plane. The following combinations are therefore possible: HH, HV, VV, or VH, of which generally only three polarization returns are independent, since $HV \approx HV$. Multipolarization therefore adds another dimension to radar remote sensing.

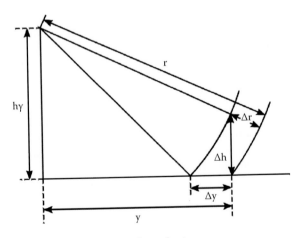

Figure 2.55   Height displacements in the radar image.

Radar Interferometry

Radar pulses are transmitted as coherent waves, and they are reflected as such. If the returning waves are received by two spatially separated antennas, then the two wave signals may be compared with respect to the phase difference by means of interferometry.

The interferometric principle is shown in Figure 2.56. The axis of the interferometer perpendicular to the base, b, yields signal differences with a phase difference of zero. At an angle to that axis, phase differences will be observed that are proportional to an angle in relation to the base and half the wavelength. This permits the location of the surface in terms of a digital elevation model.

If the phase arriving at antenna 1 (transmission and reception) is

$$\phi_1 = -\frac{4\pi}{\lambda} \cdot r$$

and the phase arriving at antenna 2 (reception only) is

$$\phi_2 = -\frac{4\pi}{\lambda}(r + \Delta r)$$

then the phase difference to be observed is

$$\Delta\phi = \phi_1 - \phi_2 = \frac{4\pi}{\lambda}\Delta r$$

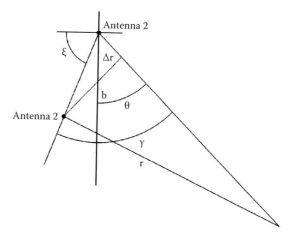

Figure 2.56   Radar interferometry.

Introducing the base, b, between the antennas, and the angles, ξ, between horizon and base as well as the angle, θ, between the point of the reflection and the vertical, one obtains

$$(r + \Delta r)^2 = r^2 + b^2 - 2rb \cos \gamma$$

with $\gamma = 90° - \xi + \theta$

Thus:

$$\Delta r = \sqrt{r^2 + b^2 + 2rb \sin(\theta - \xi)} - r^2 \approx b \sin(\theta - \xi)$$

The distance, y, from the flight axis, where a phase difference of 0 occurs is $y = r \sin \theta$. The next 0 phase difference occurs at location $y = r \sin(\theta - \xi)$. The phase differences can be made visible in the form of interferometric fringes.

Figure 2.57 shows an ERS radar image for the German island of Rügen. Figure 2.58 shows the generated interferometric fringes for two radar images. Figure 2.59 shows the derived digital elevation model for the area of Hannover derived from two subsequent ERS images.

For airborne radars, a second antenna may easily be accommodated on the aircraft. For satellites this is more difficult. During the Shuttle Radar Topographic Mission (SRTM) flown by NASA and the German DLR on the space shuttle, the second reception antenna was placed on a 60 m long beam extended from the shuttle during the radar-mapping mission. The spatial position of the two antennas and the length of the base were determined by GPS receivers.

It is also possible to create interferograms from radar signals of two different satellites, as was done during the European Space Agency's ERS1/ERS2 Tandem Mission, in which the second satellite, ERS2, followed the first, ERS1, one day later in the same orbit. On the assumption that orbital differences created a small base of between 100 to 300 m, and that the radar reflection properties have not changed during one day, interferometry became possible. ERS1 and ERS2, however, did not have precise orbital data, so that the length and the orientation of the base had to be estimated. This made the interferometric fringes ambiguous, and a restitution of the interferogram into digital elevation models required a trial and error "phase unwrapping" procedure.

For interferograms produced from the ERS 1/2 Tandem Mission, the agreement with a precise digital elevation model was within 5 m in open flat areas, although it was considerably less in forested areas, and it reached 100 m in mountain areas due to radar shadows and foreshortening.

The TerraSAR-X satellite of DLR, Germany-Astrium and the Italian-French COSMO-Skymed radar satellite succeeded to provide imagery up to 1 m GSD

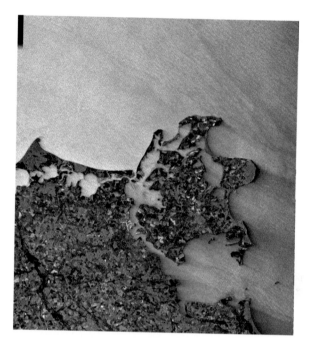

Figure 2.57 Radar image of the island of Rügen, Germany. (ERS-1 SAR © ESA, processed by DLR, courtesy of DLR, Oberpfaffenhofen, Germany.)

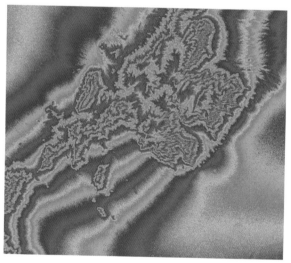

Figure 2.58 Interferogram of two radar images. (DEM from ERS 1/2 SAR © ESA, courtesy of DLR, Oberpfaffenhofen, Germany.)

Figure 2.59 Digital elevation model derived from interferogram for the area of Hannover, Germany. (DEM from ERS 1/2 SAR © ESA, courtesy of DLR, Oberpfaffenhofen, Germany.)

in X-band. A TerraSAR-X image for the Swiss city of Solothurn is shown in Figure 2.60. Another image (Figure 2.61) demonstrates the distortions of radar geometry imaging a highway bridge and hilly terrain. Figure 2.62 shows an open pit coal mine near Cologne, Germany. Figure 2.63 illustrates the Tandem-X configuration together with TerraSAR-X to generate interferometric data for elevation models. Figure 2.64 gives the result of an interferometric survey of the Hamburg harbor in Germany, showing elevation changes in color due to subsidence.

Radar Imaging Modes

Satellite radar images from high-resolution satellites, such as TerraSAR-X or Radarsat can be generated by three different scan modes:

- The basic imaging mode is the Stripmap SAR. In this mode, the antenna orientation is fixed and the side-looking radar image is generated line by line, as the satellite orbits.
- In the Scan SAR mode, a wide area of coverage is achieved by simultaneously scanning adjacent subswaths, generating lower-resolution overview images.
- In the Spotlight mode, a higher geometric resolution is obtained by directing the radar beam to the same area by multiple pulses over time.

For TerraSAR-X, the GSD in Scan SAR is 16 m, in Stripmap it is SAR 3 m, and in Spotlight it is mode 1 m or better.

Figure 2.60 TerraSAR-X image of the city Solothurn, Switzerland. (From DLR, Oberpfaffenhofen, Germany.)

Figure 2.61 Radar imaging distortions. (From DLR, Oberpfaffenhofen, Germany.)

Figure 2.62   TerraSAR-X image of coal surface mining near Cologne, Germany. (From DLR, Oberpfaffenhofen, Germany.)

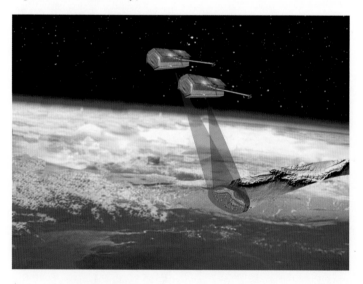

Figure 2.63   Tandem X–TerraSAR Interferometry. (From DLR, Oberpfaffenhofen, Germany.)

Figure 2.64 Interferometric survey of elevation changes in Hamburg harbor, Germany. (From DLR, Oberpfaffenhofen, Germany.)

PLATFORMS

Aircraft

The classic sensor platform is the aircraft. In order to systematically cover a portion of the earth's surface by aerial photography, flight planning is required. The aerial flights are arranged in parallel strips allowing a sufficient overlap of imaged areas by about 20% to 30%. Along the flight axis, an overlap of 60% is generally chosen, so that an overlapping pair of photos may permit the location of any photo point in at least two photographs. Since two photos are required from different exposure stations to determine an object point in three dimensions, the overlap scheme not only fulfills this condition but also permits the construction of a strong geometric interconnection between the adjacent images. The flight-planning scheme is shown in Figure 2.65.

An aerial photograph has the standardized dimension, a', of 23 cm. Thus, the area covered by a vertical photograph $A = a^2$, in which

$$a = \frac{h}{f} a'$$

with the flying height, h, and the principal distance, f.

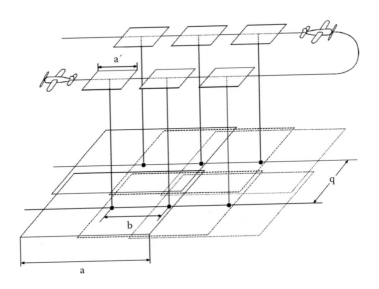

Figure 2.65   Flight plan.

The base between two subsequent photos in the strip becomes

$$b = a\left(1 - \frac{o}{100}\right)$$

in which o is the chosen longitudinal overlap of 60%. Thus, b is usually:

$$b = a \cdot 0.4$$

When the overlap between strips, p, is 20%, then the distance between two strips, q, becomes

$$q = a\left(1 - \frac{p}{100}\right) = a \cdot 0.8$$

A single model composed of the interior parts of the two photographs is of significance to calculate the number of photos required. This neat area, N, has the dimensions:

$$N = b \cdot q$$

Thus, the number of photos, n, required to cover a total area, B, is

$$n = \frac{B}{N}$$

For these photos, a film length of 0.25 m × n will be required. The scale of the image is 1:scale factor. The image scale factor is given by the ratio h/f. h is the flying height above ground. To reach the required overlap conditions, this height is always taken at the maximum altitudes of the terrain to be imaged.

Among the possible lens cones to be chosen for the flight are wide-angle objectives with the principal distance of 153 mm or normal angle objectives with the principal distance of 305 mm. In special cases, super wide-angle objectives with a principal distance of 85 to 88 mm may be utilized.

Wide-angle objectives are the most widely used. They offer about equal accuracy to determine positions and elevations. However, they have larger displacements by elevated objects such as buildings and trees. Thus, larger areas may become hidden through these objects. For this reason, urban and forest surveys in general prefer normal angle photography.

The aircraft altitude depends on the nature of the aircraft. Only military planes reach altitudes between 12 and 25 km. From 8 to 12 km, jet aircraft are required. Below 8 km, turbo prop and propeller-driven airplanes may be used. The lowest altitude, h_{min}, at which aerial flights can be made depends on the flight velocity with respect to the ground, v, and the required minimum time interval between exposures fixed by the camera, Δt_{min}:

$$h_{min} = \frac{v}{a'\left(1-\dfrac{0}{100}\right)} \cdot \Delta t_{min}$$

As a rule, aerial flights are only made in clear, cloudless weather conditions when the solar altitude is higher than 30%, but it should not exceed 60°. This limits the flying season for certain areas of the globe. For topographic surveys, a winter flying season is preferred, when the foliage of trees is minimal. Central Europe, in general, only has about 22 to 28 days per year for which aerial surveys can be made.

Aerial flight navigation before the advent of GPS required the use of a variety of electronic navigation systems. Today, automatic GPS navigation with devices offered by the camera manufacturers and their affiliates have become the rule (CCNS by IGI, T-Flight and POS Z/I by Z/I Imaging, Ascot by LH Systems).

Two companies, Applanix of Canada and IGI of Germany, offer additional inertial devices, which permit the recording of flight attitudes in three axes of rotation, in addition to determining the coordinates of the exposure station at the time of exposure.

While relative positioning via differential GPS (DGPS) is possible to about ±10 to 15 cm, angular parameters of ±0.003° may be obtained for pitch (φ) and roll (ω) and of ±0.007° for yaw (κ) by inertial devices. This is achieved by accelerometers for which the signals are integrated in an inertial measuring unit (IMU). Boresight calibration procedures are required before the flight to resolve the transformation between the three spatial coordinate systems for IMU, GPS, and camera.

It has been shown that satisfactory operations are possible to cover large areas in high altitude, small-scale flights (<1:30000) without the use of ground control and the need for an aerotriangulation. This is particularly useful for large orthophoto projects. Also at larger photographic scales, GPS positioning and IMU attitude data can be input into aerial triangulation block adjustments to minimize the required ground control.

Satellites

After the launch of the first satellite Sputnik by the former USSR in 1957, the first U.S. satellites, such as Tiros 1 in 1960, began to carry remote sensing devices to image weather patterns.

An undisturbed satellite can stay in circular orbit, if its velocity, v, is chosen in accordance with the mass, M, around which the satellite orbits, the gravitational constant, G, and the radius of the orbit from the center of mass, r:

$$v = \sqrt{\frac{G \cdot M}{r}}$$

GM for the earth is a constant $3.980 \cdot 10^5 \, \text{km}^3/\text{sec}^2$,

$$r = r_o + h$$

with $r_o = 6370$ km and h is the satellite altitude above the surface.

The period of one revolution, U, in minutes of time, as derived from Kepler's third law, is:

$$U = 84.491 \cdot \sqrt{\frac{r^3}{r^3_o}}$$

This means that a great number of satellites can be kept in orbit, as shown in Table 2.8.

Most earth-orbiting satellites have near-circular orbits for which these simplified relations are valid.

Another orbital characteristic is the inclination, I, expressed as an angle between the orbital plane and the equatorial plane, and its relation to the vernal equinox.

Geostationary satellites orbit in the equatorial plane of the earth at a speed equivalent to the earth's rotation. These satellite orbits are ideal for

TABLE 2.8   SATELLITE CHARACTERISTICS

R	h	v	U	Remarks
6700 km	330 km	7.71 km/s	90.97 min	Space station
7370 km	1000 km	7.34 km/s	105.6 min	Earth observation satellite
26570 km	20200 km	3.87 km/s	12 h	GPS
42160 km	35790 km	3.07 km/s	23 h 56 m	Geostationary satellite
384400 km	—	1.02 km/s	27 d 08 m	Moon

communication satellites and for global weather satellites looking at the entire hemisphere (Meteosat, GOES1 and 2, GMS, and Insat). Their longitude positions are at 0°, 75°W, 135°W, 140°E, and 74°E.

Most Earth observation satellites (Landsat, Spot, etc.) prefer imaging in the mode of a sun-synchronous satellite. This is possible when a constant relation between orbital node and the direction to the sun is maintained.

The orbit with its inclination and the earth rotation determine the ground track of the satellite. Repetition of that track after a specified number of days is advantageous for earth observation satellites, permitting the gathering of images within a predetermined pattern.

A satellite is subjected to various orbit-degrading influences (e.g., solar drag). To keep it in the predetermined orbit, fuel must be carried on board. It is used to make occasional thrust maneuvers to achieve the required orbit corrections.

Ground Resolution versus Repeatability

Figure 2.66 illustrates the choice of a remote sensing system with its achievable resolution and with the repeatability to obtain data.

Geostationary meteorological satellites such as Meteosat and GOES permit the imaging of the entire hemisphere, as seen from an altitude of 35790 km, by an electromechanical scanning radiometer kept in constant rotation with 5 km pixels in three bands, one of which is thermal. This can be achieved every 30 minutes. Meteosat images are placed on the Internet at least three times a day.

Geostationary satellites permit the study of weather patterns throughout the day, viewing moving clouds on TV, but they are too coarse in resolution for vegetation studies and for viewing the polar regions. Thus, NOAA satellites with a polar orbit and an altitude of about 850 km permit the imaging of the earth every 12 hours at 1 km pixels at a swath of 2700 km in five channels.

Sun-synchronous remote sensing satellites such as Landsat (USA), Spot (France), and IRS (India) have a repetition rate of less than a month, but due to high cloud cover probability an image may be obtained only several times per year at medium resolutions between 5 and 15 m in panchromatic and

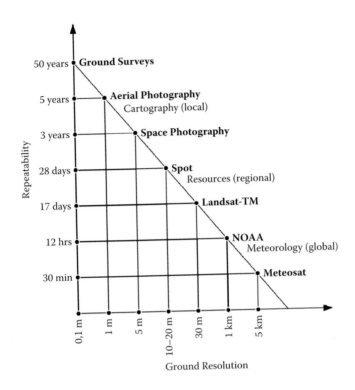

Figure 2.66 Ground resolution versus repeatability.

between 20 and 30 m in multispectral mode at swaths between 185 to 60 km.

High-resolution systems from space, such as Ikonos 2 with 1 m pixel and Russian space photography at 2 m resolution, attempt to compete with high-altitude aerial photography. However, a high area coverage with repetitivity of less than every few years is not possible.

The highest accuracy in the centimeter to decimeter range is achieved by low-altitude aerial surveys and by ground surveys. Whether these methods can be utilized over large areas is dependent on questions of cost and time. It is appropriate to apply low-resolution, high-repetitivity sensors for global surveys, intermediate resolution and intermediate repetitivity surveys, and high-resolution and low-repetitivity sensors for local (e.g., urban) surveys.

The space age began with the launch of the first satellite, Sputnik, in 1957. During the subsequent 55 years, satellite technology rapidly developed and matured. In 1961, Yuri Gagarin from Russia became the first man in space. In 1969, Neil Armstrong from the United States made his first steps on the moon. In 1981, the U.S. Space Shuttle became the first returnable platform to bring man into space and back. This was followed in 1986 by the first space platform, MIR, leading the way to the International Space Station.

In the meantime, space technology gave rise to three phenomenal developments. Depending on the launch capabilities, three types of satellites have been introduced into practice:

1. Earth observations from low-orbiting satellites, imaging the earth with a variety of sensors at high spatial resolution, with the ability to cover changing phenomena of the earth's environment in comparison to aerial imaging platforms.
2. Medium altitude satellites in near-geostationary orbits to serve telecommunication and navigational needs.
3. Space exploration for extraterrestrial missions. This has led to an otherwise not achievable understanding of our solar system. Between 1957 and 2012, altogether 75 missions were sent to the moon, 43 to Venus, 43 to Mars, 9 to Jupiter, 5 to Saturn, 2 to Mercury, 1 to Uranus, 1 to Neptune, and 12 to the Sun.

In the context of this book, earth observations by low-orbiting satellites observing the earth are of prime interest. Tables 2.11 to 2.15 demonstrate the enormous international progress made during the last 12 to 14 years. Examples of recent images are shown in Figures 2.67 to 2.72.

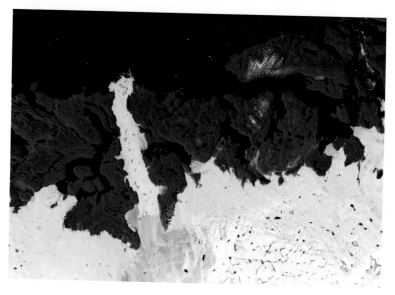

Figure 2.67 Landsat TM–ETM image of Greenland Jacobshavn glacier. (From DLR, Oberpfaffenhofen, Germany.)

Figure 2.68 Ikonos image over Cairo, Egypt. (From DLR, Oberpfaffenhofen, Germany.)

Figure 2.69 GeoEye1 image over Cologne, Germany. (From DLR, Oberpfaffenhofen, Germany.)

Figure 2.70   WorldView2 image over Dubai, UAE. (From DLR, Oberpfaffenhofen, Germany.)

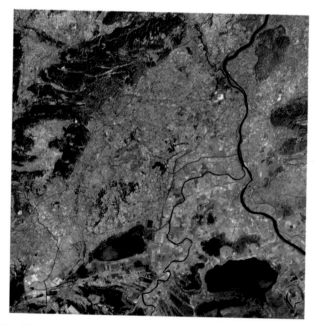

Figure 2.71   Chinese ZY-3 pan image of Rhone delta, France. (From LIESMARS, Wuhan University, China.)

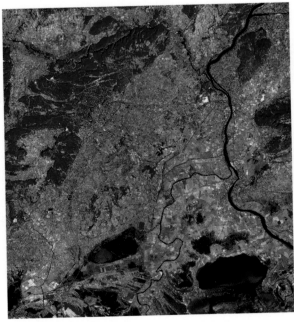

Figure 2.72 Chinese ZY-3 MS image of Rhone delta, France. (From LIESMARS, Wuhan University, China.)

With respect to earth observations the following tendencies are noted:

- A multitude of sensors have now been introduced to monitor the physical behavior of our planet. The European Space Agency's (ESA's) Copernicus program has been introduced for a systematic monitoring of environmental changes.
- High-resolution optical sensors flying in constellations are now capable of monitoring rapidly changing phenomena on the earth's surface (e.g., RapidEye).
- Time series observations secured in databases are now capable of analyzing long-term environmental changes of the earth's atmosphere and the earth's surface (e.g., Landsat 1 to 8).
- The efforts to develop small satellites with miniaturized sensor and infrastructure components have made it possible to expand the possibilities, previously only available to powerful space nations, to multinational opportunities open to developing countries.
- With the great number of new satellites planned, these tendencies are expected to continue in the future.
- One of the drivers of this development is disaster management. From 1994 to 2003, there were 6145 disasters reported around the globe.

Seventy-five percent had hydrometeorological causes, 16% were biological threats, and 8% were geological events (earthquakes). In the year 2008 alone there were 397 disasters causing the deaths of 235,000 people and a loss of US$190 billion in damages.

To effectively lessen the impact of such disasters a chain of action is required. This begins with risk assessment followed by prediction and the monitoring of actual disasters, disseminating emergency information, and response and recovery action.

Earth observation technologies, satellite communication, and satellite navigation are important elements in these tasks. A remarkable voluntary program is the International Charter for Disasters program agreed upon by the international space agencies around the globe under the United Nations Office of Outer Space to provide rapid remote sensing information free of charge to the disaster areas.

Unmanned Aerial Vehicles (UAVs)

UAVs are navigated and operated by steering controls from the ground without an onboard flight crew.

Such systems have been developed mainly for military interests (e.g., the RQ-4B Northrop Grumman Global Hawk or the Air Robot AR 100-B in the United States). Such UAVs can carry lightweight digital cameras. With these, aerial photogrammetric imagery may be recorded and restituted after recovery or transmission of the data by methods used with digital aerial mapping cameras.

Recently, UAVs have become available for civilian operators at an affordable cost. However, the legislation for the use of UAVs in different countries of the globe is still vague. Nevertheless, successful and cost-effective applications have been demonstrated for local areas and for special purposes (e.g., archaeology).

IMAGE INTERPRETATION

The image generated by a remote sensing sensor is subject to interpretation, before the remote sensing data can become information. Although there is research looking at an automation process for the information extraction procedure, currently all practical interpretations are based on the human eye–brain system.

The Human Eye

The eye performs the task of optical imaging, while the brain performs the analysis of the perceived optical data. Figure 2.73 describes the composition of the human eye.

The eye possesses a lens that can change its curvature for focusing a near or far object onto the retina, creating an image. This change of focus is achieved by the movement of a muscle. The incoming light intensity on the retina is

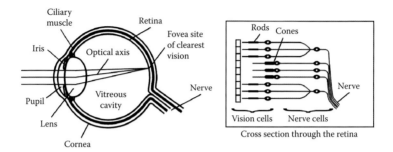

Figure 2.73   The human eye.

regulated by the pigmented part of the eye controlling the variable aperture (the pupil). The part of the retina with the highest concentration of light-sensitive cells is the fovea, extending 1° to 2° from the optical axis of the eye, whereas peripheral, low-resolution vision is possible in a range of 120°.

The young eye can change its focal length between 17 and 23 mm. The attainable resolution in the fovea is about 71 p/mm. An aerial photo of 401 p/mm can thus be observed at five- to sixfold magnification by a lens system.

The light-sensitive elements are the more densely spaced rods suitable for panchromatic sensing at low energy levels. The more widely spaced cones permit the observation of spectrally mixed energy in the form of color. The eye contains about 10^8 rods and 10^7 cones.

The lens system of the eye projects an inverted image onto the retina. The erect perception of the image is a function of the brain: the rods and cones cause chemically produced signals transmitted to the visual area of the cortex of the brain for signal processing by neural networks.

This processing analyzes the images received for gray level, color, texture, size, and context and motion, and converts them into information through a comparison with stored information in the neurons (nerve cells) of the brain. With about 10^{11} neurons contained in the brain, visual perception by far exceeds the image analysis capabilities of a computer.

Stereovision

Another interpretation tool exists in human vision; two eyes permit the fusion of two images taken from spatially different observation points, allowing a judgment of the distance of the observed object, y. Figure 2.74 shows the capacity for natural stereoscopic vision.

The human eye base, b_E, is about 65 mm. Natural stereoscopic vision diminishes with the distance squared. The stereoscopic observation capability of the brain is $d\gamma = 15''$. Thus, the ability of the brain to judge distance differences, dy, at a distance, y, is listed in Table 2.9.

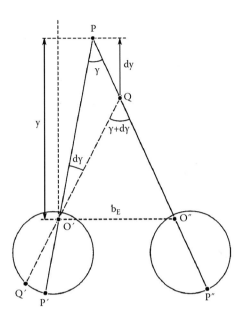

Figure 2.74   Natural stereoscopic vision.

γ, which is a small angle,

$$\gamma \approx \frac{b_E}{y}$$

By differentiation

$$d\gamma = -\frac{b_E}{y^2} dy$$

The ability to fuse images is furthermore limited by the angular range between farthest object, γ_F, and closed object, γ_c:

$$\gamma_F - \gamma_c \leq 70'$$

Image interpretation and photogrammetry have the possibility to expand the stereoscopic observation capacity to judge and to measure distances stereoscopically through the use of images, which have been taken at $n\times$ magnification of

TABLE 2.9	NATURAL STEREOSCOPIC DEPTH PERCEPTION LIMIT			
y	0.25 m	10 m	100 m	500 m	894 m
dy	0.07 mm	0.1 m	11 m	280 m	894 m

the eye base. If the two images are presented to both eyes with a magnification, *m*, and at a magnification of the eye base

$$n = \frac{b}{b_E}$$

then

$$dy = -\frac{1}{n \cdot m} \cdot \frac{y^2}{b_E} \cdot d\gamma = -\frac{1}{m} \cdot \frac{y^2}{b} \cdot d\gamma$$

Two aerial photographs taken from a distance (flying height) of 1000 m, viewed at a 4× enlargement and at 60% overlap with a base, *b*, of 400 m thus permit a stereoscopic height determination from the images at 0.05 m.

For the observation of stereo adjacent aerial photographs, it is necessary to orient the images according to epipolar rays (see Figure 2.75). Epipolar rays

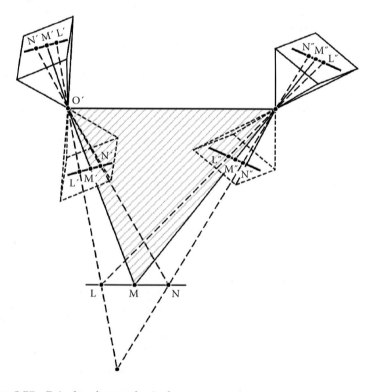

Figure 2.75 Epipolar plane and epipolar rays.

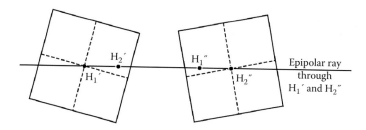

Figure 2.76   Orientation of aerial photographs according to epipolar rays.

are the lines of projection with the plane formed by the object point and the projection centers of the two images (see Figure 2.76). In practice, it is helpful to transfer the principal points H'_1 from left to right H''_1 and from right H''_2 to left image H'_2 and to position the images along a straight line containing H'_1, H'_2, H''_1 and H''_2 (Figure 2.76). Along this line (and parallel to it) the image points L'M'N' and L''M''N'' can be viewed in stereo as points L, M, N (Figure 2.75). The brain is able to compensate for minor differences in that direction of about 2%, as well as scale differences up to 5%.

The images are easiest observed with lens stereoscopes having an eye base of 65 mm and a lens magnification of 1.6. The images placed along the epipolar line then require a separation of 65 mm (see Figure 2.77).

Aerial photographs of the size 23 × 23 m are better observed in a mirror stereoscope, which separates the images by a mirror system and allows a magnification of up to four times (see Figure 2.78).

Stereo observation is also possible via anaglyphs in complementary colors (red and green) when viewed through corresponding filters. The anaglyphic images are projected or printed on top of each other in the respective

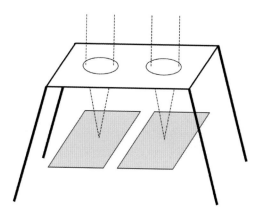

Figure 2.77   Lens stereoscope.

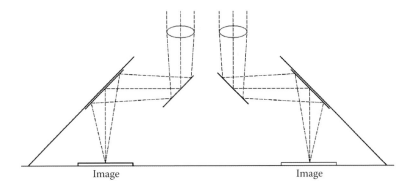

Figure 2.78   Mirror stereoscope.

colors, and viewing through filtered spectacles is possible without lenses (Figure 2.79).

Color images may be viewed if they are projected in two polarizations and viewed with corresponding polarization filters. To obtain a polarized image, not only the projected light onto a projection plane needs to

Figure 2.79   Anaglyphic stereo view of two overlapping aerial photos of Louisville, Kentucky. (Data provider: Photo Science, Inc.; Illustration provider: ERDAS Inc., the Geographic Imaging unit within Leica Geosystems GIS & Mapping Division, Atlanta, Georgia.)

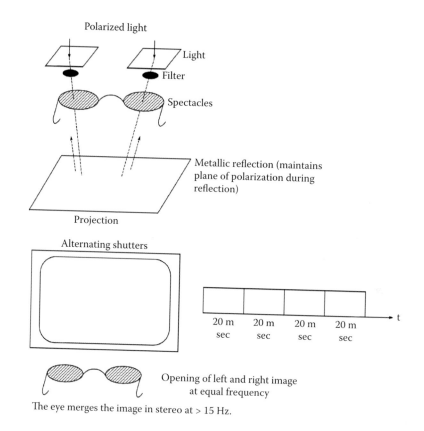

Figure 2.80 Stereo viewing by polarized light and alternating shutters.

be polarized but also the projection surface needs to reflect the projected light in polarized mode. This is possible for metallic projection surfaces (see Figure 2.80).

Stereo viewing on a computer screen becomes possible by split screens when viewed through a stereoscope. Anaglyphs may also be viewed on the screen. Special screens permit viewing with polarized light. As a rule, the "crystal eyes" principle is used, in which the left and right images are alternately generated at a 50 Hz rate. These can be viewed with filtered spectacles, which open and close the alternating left and right view at the same 50 Hz frequency.

Stereovision greatly assists in the interpretation possibilities of objects. It is also the basis for the manual photogrammetric restitution process.

Visual Interpretation of Images

Based upon the possibilities given by the human eye–brain system, the interpretation of images by an analyst starts at a primary level observing contrasts of tone and color. At a secondary level, size, shape, and texture are compared. At the third level, pattern, height difference, and shadow aids in the interpretation. At a fourth level, the association with adjacent objects plays a role.

Image interpretation has a great number of areas of application. These include:

- Military intelligence
- Forestry
- Agriculture
- Hydrology
- Topographic mapping
- Urban analysis
- Coastal area surveys
- Archaeology

For some of these applications, interpretation keys with examples of imaged objects to be interpreted have been developed.

IMAGE PROCESSING

Raster Scanning of Photographic Images

If the remotely sensed images have not been obtained by digital sensor, but by photographic imaging, the application of automated computer analysis requires a raster digitization of these images. One of the first widely used devices for this purpose, from around 1970, was the Optronics Scanner, shown in Figure 2.81. It consisted of a rotating drum, onto which the film transparency was wrapped. While the image rotated, an electronically controlled lamp sent a flash of light through the aperture and the emulsion, and collected the energy on a photo multiplier. This analogue signal was amplified and digitally converted into gray levels for each exposed pixel. After exposure of an image line, the aperture was shifted by a step motor and the next line was digitized in the same manner. If desired, the scanning of the image could easily be combined with a photowrite mechanism activated by a laser diode. This permitted online image processing under control of a computer but, as a rule, it was sufficient to store the image on a suitable storage medium. Color images could be scanned sequentially using three-color filters only permitting the recording of blue, green, and red light.

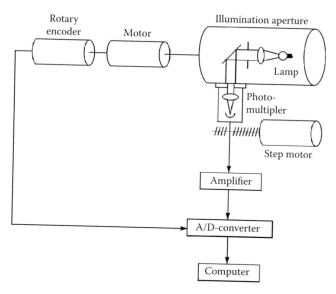

Figure 2.81   The Optronics drum scanner.

The Optronics permitted digitization in pixel increments of 12.5, 25, 50, and 100 μm. Mechanical and thermal instabilities of the instrument, however, limited the practical use to only 50 μm pixels. To reach higher digitizing resolutions, flatbed scanners have been introduced. Table 2.10 gives a summary of the available devices. Figure 2.82 demonstrates the principle of a flatbed scanner. Figure 2.83 shows the Z/I Imaging Photoscan 2001.

Gray Level Changes

Gray level changes are single pixel-based operations, which permit the changing of the available digital density of a pixel, d_i, by changing the analogue D–log H or the digital D–H curve into a new pixel density d'_i. Examples of the possibilities are shown in Figure 2.84.

- For a change of the slope of the γ curve or the response curve: $d'_i = a_o + a_1 d_i$.
- The response may also be logarithmically changed: $d'_i = \log^n d_i$.
- Densities may be grouped into preselected density ranges, by a step function.
- Finally, a gray level adaptation may be achieved by a histogram linearization, in which the surface under the histogram is divided into equal areas, defining new limits, d_i, which are to be imaged as d'_i with constant intervals. This is shown in Figure 2.84.

TABLE 2.10   FLATBED SCANNERS FOR AERIAL PHOTOGRAPHS

Manufacturer	Name	Sensor	Color	Image Pixel Size	Radiometric Range	Film
Leica	DSW 500	Camera with 2048 × 2048 pixel patches, stair stepped	Yes	Continuous 4 to 20 μm	0.1–2.5D	Single or roll film
Vexcel	Ultrascan 5000	Trilinear array	Yes	Continuous 2.5 to 2500 μm	0–4D	Single or roll film
Wehrli	RM2 Rastermaster	Linear array	No	10 μm	0.1–2D	Single film
Z/I	Photoscan 2001	Trilinear CCD	Yes	7, 14, 21, 28, 56, 112, 224 μm	0.1–3D	Single or roll film

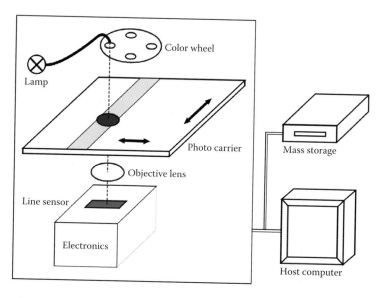

Figure 2.82   Principle of a flatbed scanner.

Figure 2.83   Z/I Photoscan 2001. (Courtesy of Z/I Imaging Corporation [Photoscan™], Huntsville, Alabama.)

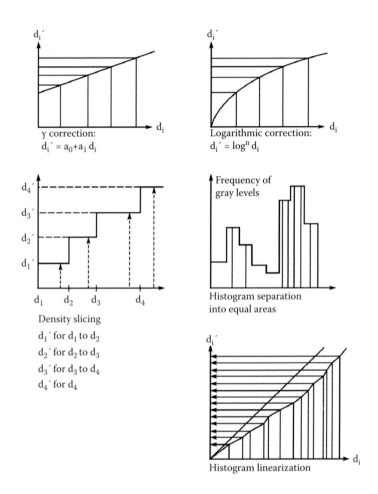

Figure 2.84   Gray level changes.

Gray level changes permit the display of an image at better contrast conditions for viewing, since the eye favors a density range between 0.1 and 1.0d.

Filtering

Filtering operations involve the adjacent pixels for each pixel. Local filters operate on an image matrix, $D(x, y)$, with a finite, usually small dimension, for example, 3×3 or 5×5. For this matrix there exists an assigned weight matrix, $W(x, y)$. The result of the filtering process for each pixel, $d(x, y)$, results out of a convolution of the image matrix, $D(x, y)$, with the weight matrix, $W(x, y)$.

IMAGE D(X,Y)

$d_{i-1,j-1}$	$d_{i,j-1}$	$d_{i+1,j-1}$
$d_{i-1,j}$	$d_{i,j}$	$d_{i+1,j}$
$d_{i-1,j+1}$	$d_{i,j+1}$	$d_{i+1,j+1}$

FILTER W(X,Y)

1/9	1/9	1/9
1/9	1/9	1/9
1/9	1/9	1/9

Low pass

+1/3	+1/3	+1/3
0	0	0
−1/3	−1/3	−1/3

Vertical direction

+1/3	0	−1/3
+1/3	0	−1/3
+1/3	0	−1/3

Horizontal direction

0	+1	+1
−1	0	+1
−1	−1	0

Diagonal direction

+1	+1	0
+1	0	−1
0	−1	−1

Diagonal direction

Figure 2.85   Low-pass filter and directional filters.

Examples for a 3 × 3 image matrix, $D(x, y)$, and a number of important weight matrices, $W(x, y)$, are shown in Figures 2.85 and 2.86.

The result of the convolution for pixel i, j is:

$$D'(i, j) = W(x, y) \,^*D(x, y)$$

- Applied to a low-pass filter this means:

$$d'_{ij} = \frac{1}{9}\left(\sum_{k=i-1}^{k=i+1}\sum_{\ell=j-1}^{\ell=j+1} d_{k,\ell}\right)$$

- Applied to a vertical directional contrast:

$$d'_{ij} = \sum_{k=i-1}^{k=i+1}\left|d_{k,j-1} - d_{k,j+1}\right|$$

- Applied to a horizontal directional contrast:

$$d'_{ij} = \sum_{k=j-1}^{k=j+1}\left|d_{i-1,k} - d_{i+1,k}\right|$$

0	−1	0
−1	4	−1
0	−1	0

Laplace edge enhancement

1	2	1
0	0	0
−1	−2	−1

Sobel edge enhancement W(x,y)

−1	0	1
−2	0	2
−1	0	1

Sobel edge enhancement W₂(x,y)

$$W = \sqrt{W_1^2 + W_1^2}$$

Figure 2.86 Edge enhancement filters.

Diagonal contrasts can be obtained in an analogous manner by the shown filters.

- The Laplace operator for edge enhancement is:

$$\Delta d(i,j) = \frac{\delta^2 d_{ij}}{\delta x^2} + \frac{\delta^2 d_{ij}}{\delta y^2}$$

Its approximation becomes:

$$d'_{ij} = 4d_{ij} - (d_{i-1,j} + d_{i,j-1} + d_{i+1,j} + d_{i,j+1})$$

- Edge enhancement is also possible by the Sobel operator, in which two filter matrices $W_1(x, y)$ and $W_2(x, y)$ are applied simultaneously.
- A high-pass filter results from the subtraction of a low-pass filtered image from the original image:

$$d'_{ij} = 2d_{ij} - d'_{i,j}$$

The filtering operations are executed pixel by pixel for the entire image.

Another type of filtering is possible for the entire image using the frequencies at which the gray values occur. This involves calculation of a Fourier transform for the image. It is basically a coordinate transformation from the image space in Cartesian coordinates (x, y) or in polar coordinates r, α into Fourier Space (u, υ), where the frequency of the signal f and α are represented as polar coordinates.

The calculation of a Fourier transform of the image, D, is:

$$FouD(u,\upsilon)=\int_{-\infty}^{+\infty}\int_{-\infty}^{+\infty}e^{2\pi\sqrt{-1}(ux+\upsilon y)}\cdot D(x,y)\,dx,dy$$

If a filter, W, is subjected to the same type of transformation

$$FouW(u,\upsilon)=\int_{-\infty}^{+\infty}\int_{-\infty}^{+\infty}e^{2\pi\sqrt{-1}(ux+\upsilon y)}\cdot W(x,y)\,dx,dy$$

then the filtered image Fourier transform can easily be obtained by the multiplication of the two Fourier transforms in u, υ space:

$$FouD'(u,\upsilon)=FouW(u,\upsilon)\cdot FouD(u,\upsilon)$$

$D'(u, v)$ can be subjected to an inverse Fourier transformation to obtain the filtered image, $D'(x, y)$, in the image space:

$$D'(x,y)=Fou^{-1}D'(u,\upsilon)=\int_{-\infty}^{+\infty}\int_{-\infty}^{+\infty}e^{-2\pi\sqrt{i}(ux+\upsilon y)}\cdot FouD'(u,\upsilon)\,du,d\upsilon$$

Use of Fourier filtering is particularly useful for the purposes of image reconstruction, when known or estimated sources of image degeneration need to be eliminated or, at least, reduced.

For the source of degradation (defocusing, image motion), a modulation transfer function $M(u, \upsilon)$ can be set up

$$\frac{1}{M(u,\upsilon)}$$

can be used as an inverse filter $I(u, \upsilon)$.

The Fourier transform of the reconstructed image then becomes

$$FouD'(u, \upsilon) = FouD(u, \upsilon) \bullet I(u, \upsilon)$$

in which $FouD(u, \upsilon)$ is the Fourier transform of the degraded image.

The inverse Fourier transform of $D'(u, \upsilon)$ then yields the reconstructed image $D(x, y)$, improved in sharpness:

$$D'(x, y) = Fou^{-1}D'(u, \upsilon)$$

Geometric Resampling

The geometry of a two-dimensional image is distorted due to the imaging geometry of the sensor, the sensor orientation, and the displacement of the three-dimensional scene when imaged into two dimensions; for these deformations of the image, models exist which are described in Chapter 3. Such a model is a function between image coordinates, $x_i'y_i'$, and object coordinates, $x_i y_i z_i$:

$$x_i' = f_1(x, y, z)$$

$$y_i' = f_2(x, y, z)$$

The inverse relations may be generated from these functions as:

$$x_i' = f_3(x', y', z)$$

$$y_i' = f_4(x', y', z)$$

With the functions $f_1, f_2, f_3,$ and f_4 known, two resampling algorithms for geometric correction of the images may be used. They are illustrated in Figures 2.87 and 2.88.

In the direct method of resampling, the coordinate $x_i y_i$ of an image pixel $x_i' y_i'$ is calculated with functions f_3 and f_4. The z_i is to be interpolated within the $x_i y_i$ pixel grid. The density $d_{x'y'}'$ is transferred to that location. All calculated points are used for further interpolation of the gray values of the output pixel matrix.

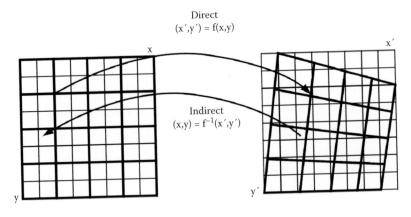

Figure 2.87   Digital rectification.

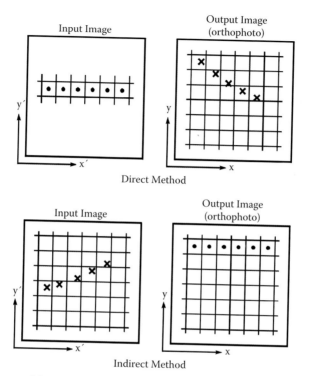

Figure 2.88   Digital rectification.

In the simpler indirect resampling method, the output pixel $x_i y_i$ coordinates with their known or interpolated height z_i permit the calculation of the location of an image point $x_i' y_i'$ assigned to the output pixel.

For this assignment, it is possible to use three options:

1. The assignment of gray values to the output pixel grid by the nearest neighbor.
2. Bilinear interpolation—The nearest four pixels for the calculated image point for the indirect method, or object pixels for the direct method, with densities of d_1 to d_4 and their distances to the output or input pixel center $x_0 y_0$ are used in a weighted function:

$$d_i = \frac{p_1 d_1 + p_2 d_2 + p_3 d_3 + p_4 d_4}{p_1 + p_3 + p_3 + p_4}$$

with

$$p_k = \frac{1}{\sqrt{(x_k - x_0)^2 + (y_k - y_0)^2}}$$

with k varying from 1 to 4.

3. Cubic convolution—Here the nearest 16 calculated points are used in the same manner.

Multispectral Classification

The objective of multispectral classification is to analyze the spectral properties of unknown objects and to compare them with spectral properties of known objects. Each spectral channel consists of a digital gray level image matrix, which geometrically coincides with the gray level image matrices of other spectral channels.

For each image a histogram of gray levels can be generated. A specific object class will produce a gray level distribution, which can be compared with a normal distribution. Statistical parameters for this comparison are, for example:

- Mean of gray levels
- Variance or the standard deviation
- Maximum and minimum gray levels for this object

Two separable objects in this channel will produce two gray level distributions.

Two channels define a two-dimensional feature space, in which the gray levels of different object types are shown as clusters (see Figure 2.89).

For n channels, there exists an n-dimensional feature space. Each pixel in any of the multispectral images can be expressed by an n-dimensional feature vector:

$$x = \begin{pmatrix} x_1 \\ x_2 \\ \vdots \\ x_n \end{pmatrix}$$

containing the gray levels in each band.

For all object clusters, a mean value, m, can be formed:

$$m = \frac{1}{K} \sum_{K=1}^{K} x_K$$

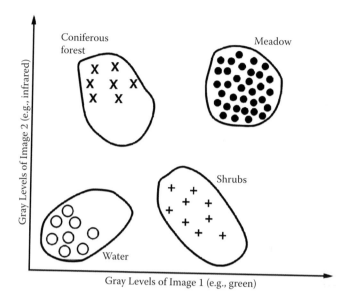

Figure 2.89   Two-dimensional feature space with object clusters.

The simplest type of classification is, then, for each group of pixels with the same geometry to form the minimum Euclidian distance to the cluster centers, m. Each pixel, i, obtains the classification of the closest cluster:

$$d(x_i, x_m) = \sqrt{(x_i - x_m)^T \cdot (x_i - x_m)}$$

The minimum distance classification can be refined into a maximum likelihood classification by use of covariance matrices for each cluster.

The covariance matrix, Σ_x, can be expressed as:

$$\Sigma_x = \frac{1}{K-1} \sum_{K=1}^{K} (x_K - m)(x_K - m)^T$$

From the covariance matrix, Σ_x, the correlation matrix, R, can be formed, in which the coefficients of the matrix are scaled down to a diagonal value of 1, so that the elements of the covariance matrix, υ_{ij}, are transformed to

$$r_{ij} = \frac{\upsilon_{ij}}{\sqrt{\upsilon_{ii} \cdot \upsilon_{jj}}}$$

The study of the covariance matrix or of the correlation matrix therefore permits the checking of the separability of the chosen object classes.

Each feature vector finds the probability of belonging to a certain class, ω:

$$p(x) = \frac{1}{(2\pi)^{n/2} |\Sigma|^{1/2}} \exp\left\{-\frac{1}{2}(x-m)^T(x-m)\right\}$$

The measure of separability between two probability distributions of the classes ω_i and ω_j is the divergence, d_{ij}:

$$d_{ij} = \int_x \left\{p(x|\omega_i) - p(x|\omega_j)\right\} \ln \frac{p(x|\omega_i)}{p(x|\omega_j)} dx$$

It represents a covariance-weighted distance between the means of two object pairs. It may be calculated as

$$d_{ij} = \frac{1}{2}T_r\left\{\left(\Sigma_i - \Sigma_j\right)\left(\Sigma_i^{-1} - \Sigma_j^{-1}\right)\right\} + \frac{1}{2}T_r\left\{\left(\Sigma_i^{-1} - \Sigma_j^{-1}\right)(m_i - m_j)(m_i - m_j)^T\right\}$$

with T_r being the trace of the matrix in question.

More refined judgments are possible by the Jeffries-Matusita distance J_{ij}, which is the distance between a pair of probability distributions:

$$J_{ij} = \int_x \left\{\sqrt{p(x|\omega_i)} - \sqrt{p(x|\omega_j)}\right\}^2 dx$$

For normally distributed classes, it becomes the Bhattacharyya distance, B:

$$B = \frac{1}{8}(m_i - m_j)^T\left\{\frac{\Sigma_i + \Sigma_j}{2}\right\}^{-1}(m_i - m_j) + \frac{1}{2}\ln\left\{\frac{|(\Sigma_i + \Sigma_j)/2|}{|\Sigma_i|^{\frac{1}{2}} \cdot |\Sigma_j|^{\frac{1}{2}}}\right\}$$

The covariance matrix, Σ_x, or the correlation matrix, R, shows that the feature vectors, x, assigned to object clusters, m, are often highly correlated between the n channels available. This permits the rotation of the feature space, x, by a rotation matrix, G, into a new feature vector space, y, so that $y = Gx$ for which the covariance matrix, Σ_y, becomes a diagonal matrix of eigenvalues, λ_i:

$$\Sigma_y = \begin{pmatrix} \lambda_1 & 0 & & 0 \\ 0 & \lambda_2 & & 0 \\ & & \ddots & \vdots \\ 0 & 0 & \cdots & \lambda_n \end{pmatrix}$$

This transformation is called a principal component transformation. Only the equations with the largest eigenvalues suffice for an optimal separation of the chosen classes in feature space.

Classification can be performed in two ways:

1. Supervised classification—It is applied if a number of object types can be recognized in the images. This implies delineation of training areas as a subset of image pixels and a generation of clusters for these areas, determining their mean vector, m. This permits the direct use of the minimum distance classifier.

   Another simple possibility is to apply a parallelepiped classifier in which the parallelepiped dimensions are formed from the maximum and minimum gray values of the training areas for a certain object, with the risk of class overlaps.

   Most appropriate is the use of the maximum likelihood classifier for each object class with its covariance matrix.

   The decision rule that x belongs to ω_i is

   $$x \in \omega_i, \text{ if } p(\omega_i|x) > p(\omega_j|x)$$

   for all $j \neq i$.

   A somewhat simpler classification is possible by the Mahalanobis distance. In the special case, that all priority probabilities are assumed equal, the decision function becomes the Mahalanobis distance:

   $$d(x, m_i)^2 = (x - m_i)^T \Sigma^{-1}(x - m_i)$$

   The Mahalanobis classifier, like the maximum likelihood classifier, retains sensitivity to direction contrary to the minimum distance classifier.

   An internal check of the classification accuracy is possible through analysis of classifications for the training areas. There it is possible to generate a confusion matrix for all object classes, listing the total numbers of pixels in each training area and their portion classified into other object classes. Obviously, an overall check in this form should be made for data obtained in the field.

2. Unsupervised classification—If no ground information to establish training areas is available, then clustering must be started by an iterative procedure estimating the likely location of clusters for ω objects. For example, in three-dimensional space, a set of clusters may be chosen along the diagonal at equal distances. Then a preliminary minimum distance classification is made, and the mean vector of the cluster centers is formed. Then the process is iterated.

   The obtained clusters can again be checked via the divergence to decide whether some clusters should be merged. A maximum likelihood classification can follow the process. At the end the classification result can be assigned as a plausible object class.

While the statistical approach to multispectral classification prevails in practice, another approach using neural networks is possible.

In two dimensions, a straight line may be drawn between two pixels so that

$$w_1 x w_1 + w_2 x w_2 + w_3 = 0$$

with x_1, x_2 representing gray values and w_1, w_2, w_3 as weights. In n-dimensions, for n bands, the equation becomes

$$w_1 x_1 + w_2 x_2 + \cdots w_n x_n + w_{n+1} = 0$$

or

$$w^T x + w_{n+1} = 0$$

For a set of image data, these weights have to be determined by a training process with the decision rules:

$x \in class$ 1 if $w^T x + w_{n-1} > 0$, or $w^T y > 0$
$x \in class$ 2 if $w^T x + w_{n-1} < 0$, or $w^T y < 0$

The weight, w, is modified to w' with a correction increment, c:

$$w' = w + cy$$

so that

$$w'^T y = w^T y + c|y|^2$$

In practice, this is iterated until

$$w'_i = w_i + cy$$

and

$$w'_j = w_j - cy$$

The disadvantage of the pixel-based multispectral classification approach is that homogeneous objects are not treated as a unit. This can be overcome by image segmentation. Image segmentation can be implemented either by edge detection techniques or by region growing. The classification algorithms may then be applied to regions rather than to pixels. A recently developed product is the eCognition "context-based" classifier.

If segmented data are available through a GIS system, a knowledge-based classification approach may be applied to image regions. Knowledge is intro-duced by a set of rules: *if* a condition exists, *then* inference is applied. Figure 2.90 shows the example of a semantic network, which can be used to test segmented images for their content.

Examples for such rules are

- If Landsat band 7 > Landsat band 5, then vegetation
- If radar tone is dark, then smooth surface

It is also possible to incorporate texture parameters into the classification process. Characteristic texture parameters for an image region are

- Autocorrelation function
- Fourier transforms
- Gray level occurrence

An example of a multispectral classification of a multispectral image (Figure 2.91) is shown in Figure 2.92 for the urban area of Augsburg, Germany, with a Landsat image.

Classification Accuracy

Of great interest in remote sensing is the assignment of multispectral classification accuracy.

Traditionally, the accuracy has been checked by photointerpretation, assuming that this higher-resolution classification is 100% correct. This permits an error matrix to be developed for the classes chosen and the pixels or objects correctly or incorrectly determined.

The application of statistical techniques is often handicapped because of a lack of a normal distribution for the observations.

One of the tools to compare two independent classifications is Cohen's Kappa Test. If one classification is superior in accuracy (e.g., by photointerpretation)

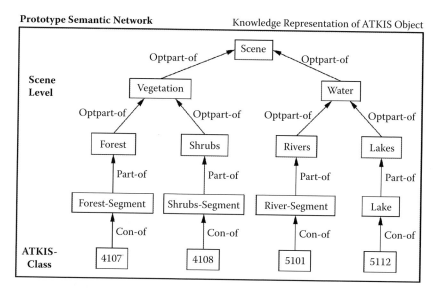

Figure 2.90 Semantic network. (From the Institute for Photogrammetry and GeoInformation, University of Hannover, Germany.)

Figure 2.91   Multispectral Landsat image of Augsburg in Germany. (Landsat 7 false color composite; USGS 2000, GAF 2000, courtesy of GAF Remote Sensing and Information Systems, Munich, Germany.)

and the other is derived from satellite imagery, the agreement for a certain object class will be less than 1 (e.g., 0.1 = poor; 0.5 = moderate; 0.7 = substantial; 0.9 = perfect).

Image Fusion

Image fusion in remote sensing is a technique by which high-resolution single-channel (panchromatic) images are geometrically merged with lower-resolution multispectral images in three spectral bands for visualization in the colors blue, green, and red. As the color space is not suitable for merging because of correlation between channels, the multispectral information is transformed into intensity, hue, and saturation of the three colors, with the intensity representing the high-resolution component of the panchromatic image.

The methodology is also applicable to more than three multispectral channels, if the number of existing bands is subjected to a principle component transformation.

Figure 2.92 Multispectral classification of Augsburg in Germany. (Land cover classification, courtesy of GAF Remote Sensing and Information Systems, Munich, Germany.)

Provided that the images are geometrically registered with respect to each other, multitemporal and multisensor data fusion may also be used. A comparison of multitemporal data is generally done for change detection. Here again, a knowledge-based approach is advantageous.

REMOTE SENSING APPLICATIONS

A great number of satellite systems have provided satellite imagery for remote sensing applications in different disciplines. Table 2.11 provides a summary of the classical remote satellite systems with optical sensors. Table 2.12 augments this list with the high repletion satellites, which have been available since 1999. Table 2.13 lists important new medium resolution satellites that have been launched since the millennium. Table 2.14 shows the early classical radar satellite systems; Table 2.15 augments the list with the recent high-resolution radar systems since the millennium.

In addition to these satellites, a great number of images have been acquired through shorter duration missions by Russian photographic camera systems from Kosmos satellites and the MIR space station, for example, the Russian KFA 1000, KVR 1000, and TK 350 cameras; the German Metric camera on the space shuttle; as well as the MOMS- 2P digital sensor flown on the space shuttle and on MIR.

TABLE 2.11   COMMONLY USED CLASSICAL OPTICAL REMOTE SENSING SATELLITES

Type	Name	Orbit	Repeat Cycle	Swath	h	Resolution	Bands	Country	Application
Meteorological	Meteosat	Geostationary	30 min	Half spheric	36000 km	5 km		ESA	Meteorology, climate
	GOES-1	75°W						USA	
	GOES-2	135°W						USA	
	GSM	140°E						Japan	
	Insat							India	
	Meteor 3							Russia	
	FY2	105°E						China	
Meteorological	NOAA	Polar	12 h	2394 km	705 km	1 km	5	USA	Climate
	DMSP	Polar		2400–3000 km		3 km	0.58–21.5 μm	USA	Military
	5 satellites	Polar					6	Russia	Climate
	Meteor P						6		(TOMS, ozone)
							0.31–0.38 μm		
Earth resources	Landsat (1–3) MSS	Sun synchronous 9:30 at 40°	18 d	185 km	918 km	80 m	3–4	USA	1972–1984

						Resolution			
	Landsat (4–5) TM	10:30 at 40°	16 d	185 km	705 km	30 m (thermal 120 m)	7	USA	1982, 1984
	Landsat 7 TM	10.00 at 40°	16 d	185 km	705 km	15 m pan	7	USA	1999
Earth resources	Spot P 1-4	Sun synchronous	26 d	60 km	832 km	30 m MS		France	1986
						10 m pan	1		
Earth resources	Spot XS 1-4	10:00 at 40°				20 m MS	3		1993
Earth resources	JERS 1 OPS			75 km	568 km	20 m	7	Japan	1992
Earth resources	IRS 1 A, B	Sun synchronous	22 d to 24 d	141 km	904 km	MS 36.6 m	3	India	1988, 1991
	IRS 1 C, D	9:25 at equator			774 km	Pan 5.6 m	1		1995, 1997
						MS 23.5 m	3		
						WIFS 188 m	2		
Cartographic	Ikonos 2	Sun synchronous		11 km	677 km	1 m pan	1	USA	1999
	EROS A1	Sun synchronous		12 km	480 km	4 m MS	1	Israel	2000
		Sun synchronous		8 km	450 km	1.8 m pan	1	USA	2001
						0.6 m pan			
						2.4 m MS			

TABLE 2.12 RECENT OPTICAL HIGH-RESOLUTION SATELLITE SYSTEMS

Agency	Satellite	Year	GSD-pan	GSD-MS	Number of Bands	Swath	Remarks
Digital Globe, USA	Ikonos	Sept 1999	0.82 m	3.2 m	4	11.3 km	
Imagesat, Israel	EROS-A	Dec 2000	1.9 m	—	—	14 km	Pan only
Digital Globe, USA	Quickbird	Oct 2001	0.61m	2.44 m	4	16.5 km	
CNES, France	Spot 5	May 2002	2.5/5 m	5/10 m	4	60 km	
Taiwan	Formosat 2	May 2004	2 m	8m	4	24 km	
ISRO, India	Cartosat 1	May 2005	2.5 m	—	—	30 km	Pan, stereo
Jaxa, Japan	Alos	Jan 2006	2.5 m	10 m	4	70 km	Stereo
Imagesat, Israel	EROS-B	April 2006	0.7m	—	—	11 km	Pan only
Russia	Resurs DK1	June 2006	0.9 m	2.5 m	3	28.3km	
Digital Globe, USA	WorldView 1	Sept 2007	0.5 m	—	—	17.6 x 14 km pan (750000 km²/d)	
ISRO, India	Cartosat 2A,B	April 2008 (July 2010)	0.8 m	—	—	9.6 km	Pan only
Digital Globe, USA	Geo Eye 1	Sept 2008	0.41 m	1.65 m	4	12.5 km (250000 km²/d)	
Digital Globe, USA	WorldView 2	Oct 2009	0.46 m	1.8 m	8	16.4 km (1 M km²/d)	
Astrium, F & D	Pleiades1A(B)	Dec 2011 (Dec 2012)	0.7 m (0.5 m resampled)	2 m	4	20 km constellation	
China	ZY-3	Jan 2012	2.1 m	3.5 m	2	51 km	Stereo
Korea	Kompsat 3	May 2012	0.7 m	2.8 m	4	16.8 km	

TABLE 2.13   RECENT OPTICAL MEDIUM-RESOLUTION SATELLITE SYSTEMS

Agency	Satellite	Year	GSD-pan	GSD-MS	Number of Bands	Swath	Remarks
ISRO, India	IRS-1C	Dec 1995	5.8 m	23 m	4	141 km	
NASA, USA	Landsat 7+ETM	April 1999	15 m	30 (90) m	7	185 km	
NASA, USA	ASTER	Dec 1999	—	15m VNIR	1–3	60 km	
			—	30m SWIR	4–9		
			—	90m TIR	10–14		
China–Brazil	CBERS 2	Oct 2003	20 m	to 260 m	4	113 km	
Germany	RapidEye	Aug 2008		6.5 m	5	77 km	Constellation, red edge, 5 satellites
UK	DMC 2	July 2009		2 2m	3	660km	
USGS, USA	Landsat 8	May 2013	15 m	30 m (100 m Thermal)	7	185 km	

TABLE 2.14   HISTORICAL RADAR SATELLITES

Name	Year	Inclination of Orbit	Swath	h	Resolution	Polarization	Country/Agency
Seasat	1978	72°	100 km	790 km	40 m	HH 23.5 cm	USA
SIR-A	1981	50°	50 km	250 km	38 m	HH 23.5 cm	USA
SIR-B	1984	58°	40 km	225 km	25 m	HH 23.5 cm	USA
SIR-C	1994	51°	30 to 60 km	225 km	Variable 13–26 m	Multiple HH, HV, VH, VV, 23.5 cm, 58 cm, 3.1 cm	USA
ERS 1/2	1991, 1995	Polar	100 km	785 km	30 m	VV 5.7 cm	ESA
JERS 1	1992	Polar	75 km	568 km	18 m	HH 23.5 cm	Japan
Almaz	1991	Polar	50 to 100 km	350 km	15 m, variable	HH	Russia
Radarsat	Since 1995	Polar	50 to 500 km	800 km	Up to 10 m	HH 5.7 cm	Canada
Envisat (ASAR)	2002	Polar	100 km	800 km	12.5 m	HH, VV	ESA

TABLE 2.15 RECENT HIGH-RESOLUTION RADAR SATELLITE IMAGING SYSTEMS

Agency	Satellite	Year	GSD	Swath	Band
ESA	ERS-1	1991	10–30 m	100 km	C
Jaxa, Japan	JERS1	1992	18 m	75 km	C
ESA	ERS-2	1995	10–30 m	100 km	C
Canada	Radarsat 1	1995	9–100 m	50–500 km	C
NASA, USA	SRTM	2000	30 m	225 km	C
DLR, Germany	SRTM	2000	30 m	45 km	X
ESA	Envisat	2002	30–1000 m	100–405 km	C
Italy, France	COSMO-Skymed	2006	1–50 m	10–200 km	X
Canada	Radarsat 2	2006	3–50 m	20–500 km	C
Germany, Astrium	TerraSAR-X	2007	1, 3, 18 m	10, 30, 100 km	X
China	SurveyorSAR	2007	10–25 m	100–250 km	C
ISRO, India	Risat 1,(2)	2009 (2012)	3–50 m	10–240 km	C
Germany, Astrium	Tandem X	2010	1, 3, 16 m	10, 20, 100 km	X radar interferometry
China	Civilian	2012		Radar satellite	S

Figure 2.93   IRS1C/D image of Munich Airport in Germany. (IRS-PAN/LISS image: Munich; SI/Antrix/Euromap 1999, GAF 2000, courtesy of GAF Remote Sensing and Information Systems, Munich, Germany.)

An example image of the airport of Munich, Germany, taken by IRS-C is shown in Figure 2.93.

These images are available from nationally and internationally operating space agencies, such as NASA, NOAA, ESA, CNES, NASDA, ISRO, or their vending agencies (e.g., USGS, Spot Image, Space Imaging, Eurimage, DigitalGlobe). The data cost still differs greatly but it is in a process of stabilization.

Meteorological data, though reduced in quality, are available over the Internet free of charge. Original resolution meteorological images can be obtained at reproduction cost. Medium- and high-resolution images have a weak to strong commercial component, depending on the policies of the space agency maintaining the satellite system. Privately funded commercial systems charge full price, unless they are supported by large government programs.

Global applications, therefore, use low-cost, low-resolution imagery, which is easily obtainable. Regional and local applications requiring higher resolutions rely on substantial imagery purchases. There, remote sensing competes with other data acquisition methods with respect to obtainable quality, cost, and evaluation effort and time.

Project-based research applications were the easiest to be realized. The present focus of applications is to concentrate on organized data acquisition and analysis programs depending on the socioeconomic priorities to be placed on applications made possible by public or industrial funding.

We will now consider the situation in the major application areas of remote sensing.

Meteorology and Climatology

Atmospheric sciences study the different layers of the earth's atmosphere:

- Troposphere, from 0 to 20 km altitude
- Stratosphere, from 20 to 50 km altitude
- Mesosphere, from 50 to 80 km altitude
- Thermosphere, from 80 to 300 km altitude

The "weather zone" is the troposphere, which is of direct meteorological interest. However, the other zones also affect weather and climate.

Ozone

The earth's ozone shield extends from an altitude of 25 to 60 km. It absorbs or reflects most of the ultraviolet (UV) light, so that only minimal amounts of ultraviolet reach the earth's surface.

NASA has launched an ultraviolet sensor, TOMS (Total Ozone Mapping Spectrometer), which observes the ozone layer in six bands between 312.5 mm and 380 mm wavelength. From these six bands, three pairs can be formed. They determine transmission minus absorption of UV energy. This ratio is a measure of ozone concentration. TOMS was carried on the U.S. satellite Nimbus 7 from 1978 to 1993, on the Russian satellite Meteor 3 from 1991 to 1994, and on a special satellite since 1996. TOMS detected a rapidly deteriorating ozone concentration over the south polar regions in 1992 and 1993, which created great public interest. The cause could have been the eruption of the volcano Mount Pinatubo in 1991, even though aerosols produced by human activity may also have played a part. Since then, ozone measurement has been a major global remote sensing application.

The European Space Agency (ESA) launched an ozone sensor on the ERS-2 satellite in 1995. Figure 2.94 shows the global ozone concentrations on a particular day (Figure 2.94a) and a monthly average (Figure 2.94b). The observations were continued by Schiamachy on Envisat under the GOME experiment for ozone studies. Figure 2.95 shows the ozone vertical column density for the northern hemisphere and Figure 2.96 for the southern hemisphere.

Cloud Mapping

The first U.S. imaging satellite launched in April 1961 was Tiros 1, which made it possible to observe clouds. Today, geostationary satellites permit a

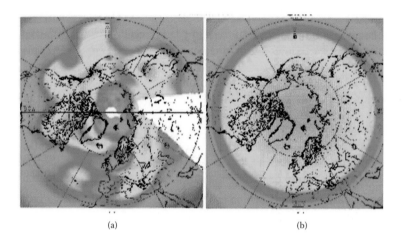

(a)   (b)

Figure 2.94   Images of ERS2 ozone sensor of the globe. (ERS-2/GOME; DLR, courtesy of DLR, Oberpfaffenhofen, Germany.)

daily weather watch following the movement of cloud patterns. The satellites Meteosat (over Africa), GOES 1 (over Venezuela), GOES 2 (over Hawaii), GMS (over the Philippines), and Insat (over the Indian Ocean) gather images every 30 minutes in the visible and thermal range. These may be geocoded and used in animations, which are commonly shown on television programs.

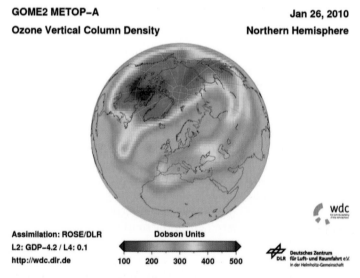

Figure 2.95   Image of Envisat Schiamachy GOME ozone sensor of the Northern Hemisphere. (From DLR, Oberpfaffenhofen, Germany.)

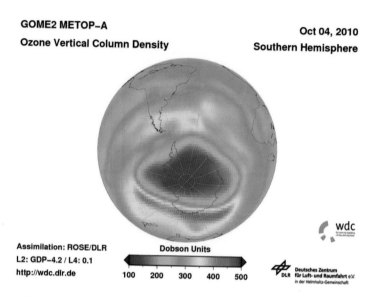

GOME2 METOP–A

Ozone Vertical Column Density

Oct 04, 2010

Southern Hemisphere

Assimilation: ROSE/DLR
L2: GDP–4.2 / L4: 0.1
http://wdc.dlr.de

Dobson Units

100   200   300   400   500

Figure 2.96   Image of Envisat Schiamachy GOME ozone sensor of the Southern Hemisphere. (From DLR, Oberpfaffenhofen, Germany.)

The data collected at an interval of 6 hours may be used to determine parameters of the radiative transfer model, into which the distribution of clouds, water, ice, snow, and the land mass is entered.

The combination of visual and thermal bands permits the visual separation of clouds, water, ice, snow, and land. If combined with atmospheric non-remote sensing measurements, a radiative transfer model can be arrived at. Figure 2.97 shows the thermal GOES-1 image of a hurricane, and Figure 2.98 gives a view of a hurricane with SeaWiFS from OrbView-2.

Rainfall

The measurement of rainfall is of great meteorological interest. The sources of worldwide rainfall data are rain gauges, which are very scarcely distributed over the globe. If no rain gauge data are available over a region, data from Meteosat, GOES, and so on relating to thermal bands can be used to determine cloud temperature. Cold clouds with <235 K temperature give an indication of possible rainfall.

On the U.S. military DMSP satellite, four wavelengths from 0.35 to 1.55 cm are provided for passive microwave sensing in two polarizations (HH, VV). The images of 55 km ground resolution permit the derivation of a scattering index, indicative of rainfall.

Figure 2.97   Thermal GOES image of hurricane. (Meteosat-3 MVISSR; Eumetsat, processed by DLR, courtesy of DLR, Oberpfaffenhofen, Germany.)

Figure 2.98   SeaWiFS OrbView-2 image of a hurricane. (From SEOS. EARSeL is a partner in the ESA project SEOS, Greece.)

Wind

Since wind drives ocean currents, scatterometers can measure the roughness of the sea to estimate wind vectors. Radar images are also able to detect roughness parameters.

Weather Prediction

Terrestrial measurements for weather forecasting can easily be combined with remote sensing data for cloud motion, the estimated precipitation, and the measurement of surface temperature.

Other phenomena detected from images are

- Analysis of snow cover
- Location and the motion of tropical storms
- Detection of fog and its dissipation

Climate Studies

Climate studies become possible by the comparison of NOAA-AVHRR aggregates on a seasonal and annual basis.

Oceanography

The geodetic aspects to be studied are ocean heights, as determined in the ESA-ERS1 Topex mission by radar altimeters. When related to the geoid, the data permit the derivation of ocean height (see Figure 2.99).

Let us now consider some phenomena of direct interest to remote sensing.

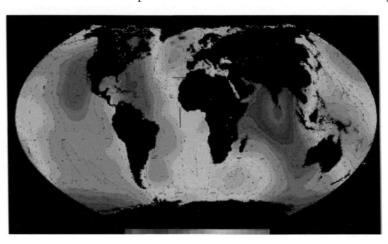

Figure 2.99 Heights from Topex altimeter ocean ERS-1. (Courtesy of DLR, Oberpfaffenhofen, Germany.)

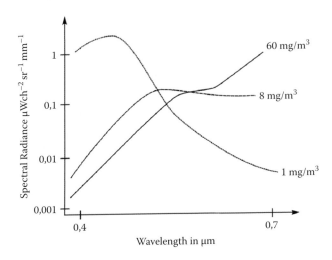

Figure 2.100   Spectral chlorophyll response at sea.

Ocean Productivity

The main objective of studying ocean productivity is to detect organic substances, such as phytoplankton, which are important for fisheries. It contains chlorophyll, which can be differentiated from suspended sediments prevalent near the coast and transported by estuaries (see Figure 2.100). Figure 2.101 shows an image of chlorophyll concentration and sediments in the Strait of Gibraltar.

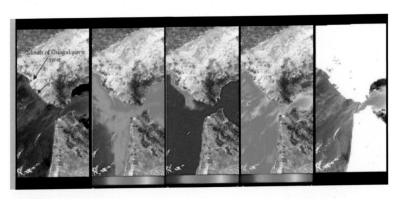

Figure 2.101   Chlorophyll separation of pigments, sediments, and aerosols from MOS. Images of IRS for the Strait of Gibraltar. (Left) Multispectral image followed by the separation of pigments (chlorophyll concentrations), sediments, and aerosol concentrations. (Right) A black-and-white image for the separation of clouds over the sea. Pigments in μg/l (blue = 0, green = 3, red = 6; sediment: blue = 0, red = 5; aerosol optical thickness: blue = 0, red = 1). (MOS-IRS; DLR, courtesy of DLR, Oberpfaffenhofen, Germany.)

Figure 2.102   Algae bloom in the Bay of Bisquay. (From SEOS. EARSeL is a partner in the ESA project SEOS, Greece.)

Sediments reflect mainly in red. Therefore, a blue/green ratio can indicate the chlorophyll concentration at sea (see Figure 2.102 for algae bloom). The actual concentration can be calibrated by *in situ* measurements.

The satellites Nimbus 7 and OrbView-2 carried the sensor SeaWiFS for observation in eight channels at 1 km resolution with a swath of 2800 km. Of particular interest is the observation of sea surface temperature, which is made available at weekly intervals (see Figure 2.103 for the eastern Mediterranean Sea).

Ocean Currents

Ocean currents are visible along the coast because of plumes of suspended matter. In the midocean, the radiant temperature, which can be measured day and night by NOAA-AVHRR, shows the distribution of ocean currents. Global thermal phenomena, like El Niño, can be monitored by NOAA satellites (see Figure 2.104).

Radar images are also an indicator that can help to determine the level of surface roughness, since currents produce small waves.

Sea Ice

The principal objective of the Canadian Radarsat satellite is to be able to follow sea ice motion in the polar areas. In multitemporal mode the images can be observed in stereo, giving an indication of the direction of flow (see Figure 2.105).

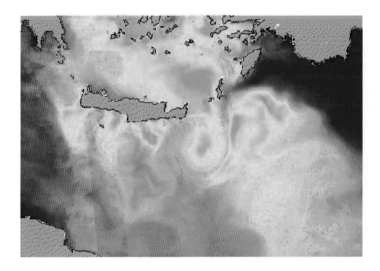

Figure 2.103   Sea surface temperature in the Eastern Mediterranean, weekly mean temperature, summer 1994. Dark red = 30°C, blue = 23°C. (From NOAA-AVHRR; DLR, courtesy of DLR, Oberpfaffenhofen, Germany.)

The ESA-ERS1/2 satellites have been able to classify Arctic sea ice. The thermal band six of Landsat TM can distinguish ice temperatures. Thin ice is warmer than thick ice.

The surface roughness of sea ice can be measured by nonimaging radar scatterometers, flown from aircraft at altitudes below 1 km.

Figure 2.104   El Niño observed sea surface temperature anomaly by ERS along track scanning radiometer (monthly mean, October 1997). (From CEOS CDROM 98, courtesy of CNES, Toulouse, France.)

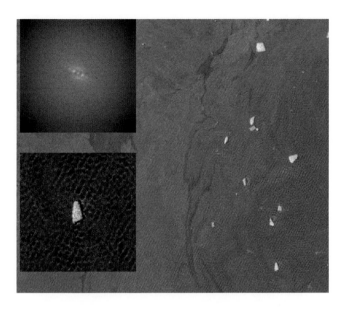

Figure 2.105   Radarsat image of icebergs on the west coast of Antarctica with wave patterns around the icebergs. (From ERS-1, © ESA, processed by DLR, courtesy of DLR, Oberpfaffenhofen, Germany.)

Bathymetry

Bathymetry at sea is generally made by sounding from ships, which have, however, difficulties in navigating in shallow areas. In these areas, remote sensing from aircraft and satellites can help to assess water depth, even though, depending on the turbidity of the water, light penetration and reflection is generally limited to not more than 10 m water depth. Figure 2.106 illustrates the transmission in water, which is best for the green band. A blue–green ratio can therefore be used for an assessment of water depth in shallow areas. Figure 2.107 shows the composition of the Wadden Sea near Wilhalmeshaven, Germany, at low tide.

Environment

Remote sensing concentrates on the detection of environmental pollution.

Hazardous Waste

The task to identify hazardous waste is to find its location, to map it, and to monitor it. If identification is not directly possible by visual inspection, the health of vegetation over the waste area can be a good indicator for hidden

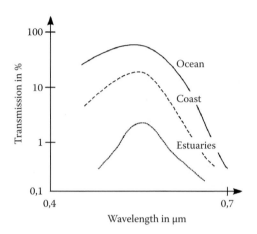

Figure 2.106   Water penetration in the visual spectrum.

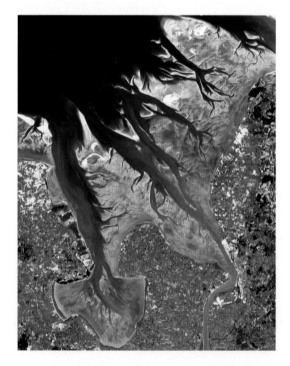

Figure 2.107   Wadden Sea on the German North Sea coast near Wilhelmshaven at low tide. (IRS 1D-LISS image, SI/Antrix/Euromap 1998, GAF 2000, courtesy of GAF Remote Sensing and Information Systems, Munich, Germany.)

waste. A ratio of infrared–red is a good discriminator. Waste areas will usually have a low infrared–red ratio.

Plumes

Thermal pollution in water can be monitored by thermal infrared scanners operated from aircraft or by multispectral optical sensors from satellites. Figure 2.108 shows the thermal pollution in the Po River, and Figure 2.109 shows the sediment concentration caused by the Po River in the Adriatic Sea.

Oil Spills

Of particular environmental interest are oil spills at sea and in coastal regions. Due to its fluorescent properties, oil floating on water has a high reflectance in

Figure 2.108 Po River thermal pollution. (Daedalus, DLR, courtesy of DLR, Oberpfaffenhofen, Germany.)

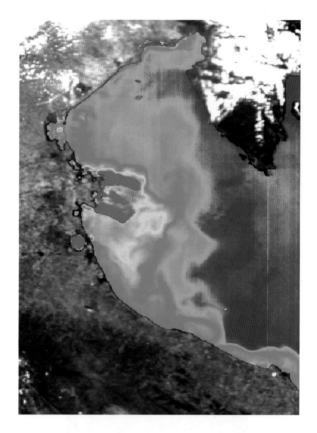

Figure 2.109 Sediment concentration at sea caused by the Po River in the Adriatic sea. (From MOS-IR; DLR, courtesy of DLR, Oberpfaffenhofen, Germany.)

the ultraviolet region (0.3 to 0.4 µm). An aircraft scanner operated from low altitude can be used to determine the thickness of the oil film on water.

On a more regional and global level, radar images permit the location of oil slicks. They dampen the natural roughness of ocean water. They are thus observable due to their low backscatter in the images. On a local level, laser scanners are able to differentiate between oil types (see Figure 2.110 as an example for a ship accident near the coast of Northern Spain). Thermal images may also be of help, since the oil surface is cooler than the water.

Nonrenewable Resources

Locating nonrenewable resources is the task of geological and geophysical exploration.

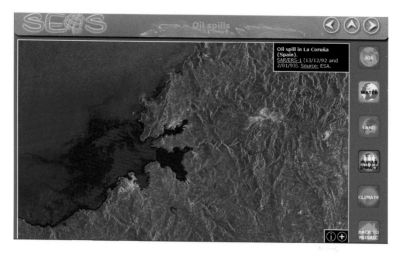

Figure 2.110   Oil pollution after ship accident on the north Spanish coast. (From SEOS. EARSeL is a partner in the ESA project SEOS, Greece.)

Mineral Exploration

For the prospecting of minerals, the methodologies of structural geology are very important. Geological prospecting is difficult in areas covered by vegetation, however, it becomes easier in the dry belts of the globe.

Ore deposits are usually associated with structure zones. The interpretation of lineaments in the rocks permits the visible determination of fractures. Digital directional filtering of the images helps to better recognize these fractures. Fractures are often also combined with hydrothermal activity. Therefore, thermal aircraft scanners are useful in this respect. To identify the mineral composition of outcrops, hyperspectral scanners, such as AVIRIS, offer special possibilities to distinguish minerals of outcrops.

Oil Exploration

Remote sensing plays an initial role in oil exploration. The analysis of Landsat and radar images to determine the extent of sedimentary basins is usually the first step in oil exploration, even though a whole slate of geophysical prospecting methods must be applied to focus on potential exploration locations. These are

- Aeromagnetic surveys
- Gravity surveys
- Reflection seismology
- Drilling

Figure 2.111   Deforestation in Bolivia. (Courtesy of SEOS, Greece.)

Renewable Resources

The task of making an inventory and then monitoring renewable resources is intimately connected with the establishment of geographic information systems. It consists of compiling base data by standard mapping procedures, of supplementing these data by geocoded remote sensing data, and of adding attribute data from a multitude of sources.

Remote sensing has the advantage of quick acquisition of up-to-date images at adapted spatial and spectral resolution without limitations of costly ground access. It has the disadvantage that not all desired categories of information can be extracted from the images. Figure 2.111 illustrates the deforestation observed from Landsat in Bolivia.

Land Cover and Land Use

Land cover describes the physical appearance of the earth's surface, whereas land use is a land right-related category of economically using the land. Remote sensing concentrates on observing land cover.

Land cover consists of classifiable terrain objects for which different governmental base data providers have implemented a number of object-oriented classification schemes. The UN Food and Agricultural Organization has devised a hierarchical land cover catalog suitable for the application of remote sensing and applicable to developing continents, such as Africa. It has been implemented in the Africover project distinguishing 90 different land cover classes in different East African countries.

The German Surveys and Mapping administration has devised an object-oriented land cover catalog with the following major categories:

1000	Text information
2000	Settlements
3000	Transportation
4000	Vegetation
5000	Hydrology and water
6000	Topographic elevation data
7000	Administrative boundaries

These major categories are subdivided into subclasses. Many of the subclasses can be identified and monitored with 85% accuracy by remote sensing at the global and regional or even at the local level. Figure 2.112 shows the result of a land cover classification from an airborne multispectral scanner. As shown in Figure 2.113, the result is significantly enhanced if it is merged with vector information from a GIS. Figure 2.114 is an example of two satellite image-based classifications for the island of Tenerife, Canary Islands, with changes of land cover.

Vegetation

Of particular significance is the monitoring of vegetation. On a global level, this is done from NOAA-AVHRR images obtainable twice a day. Many of these images contain clouds, which have to be eliminated from the images. This is possible by taking all of the images of a 10- or 14-day period and substituting cloudless pixels into the data set.

Whereas the original 1 km resolution provides too much data, a reduced resolution of 4 km or 16 km can provide a global vegetation index data set. This is possible by the use of the ratio for the Normalized Difference Vegetation Index (NDVI):

$$NDVI = \frac{Infrared - Red}{Infrared + Red}$$

The calculated NDVI for each pixel can attain a value between 1 and −1.

Green vegetation has an NDVI of about +0.7 (shown in red in Figure 2.115), whereas water, barren lands, and clouds have an NDVI of about −0.3, shown in blue.

After radiometric calibration and the consideration of atmospheric scattering and geocoding are applied, the NDVI gives a clear indication of seasonal vegetation changes. On the Northern Hemisphere, greenness tends to rise in May, with a peak in July and a decrease until September. Yearly comparisons for the respective months can be useful for comparing crop estimates from year to year.

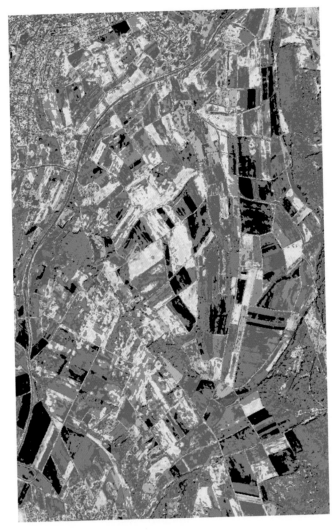

Figure 2.112 Daedalus aircraft scanner image over agricultural area. (Daedalus, © DLR, courtesy of DLR, Oberpfaffenhofen, Germany.)

Figure 2.113   Parcel-based land cover from DaedalusScanners image merged with GIS vector data. (Daedalus; DLR, courtesy of DLR, Oberpfaffenhofen, Germany.)

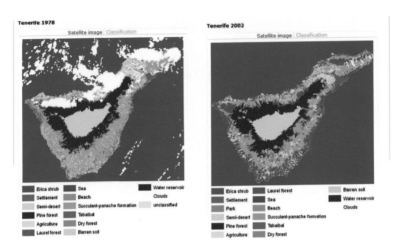

Figure 2.114 Land cover changes of Tenerife from supervised classification. (From SEOS. EARSeL is a partner in the ESA project SEOS, Greece.)

Another use of the NDVI is in the monitoring of tropical vegetation and the depletion of forests, or of agricultural crops (Figure 2.116 shows the application for a multispectral aircraft scanner in agriculture).

Natural Hazards

Earthquakes

Seismic risks are not directly observable by remote sensing. However, active faults may be visible in Landsat images, and the plate movements along these faults can be monitored by radar interferograms.

Landslides

Fresh landslides are observable in radar images.

Land Subsidence

Land subsidence can be interpreted from images by a change in drainage patterns and an observation of vegetation anomalies.

Volcanoes

Volcanic eruptions are associated with clouds of ash, slope changes, and mud-flows. Changes in topography can be monitored by radar interferometry, but it is also possible to use thermal images from Landsat TM and from aerial scanners to study changes in heat emission.

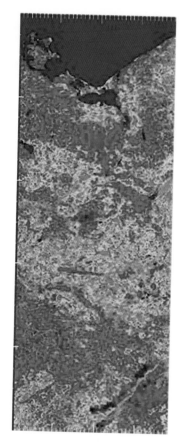

Figure 2.115   NDVI of the Oder region from MOS on IRS from channels 7 (651 nm) and 11 (869 nm). (MOS-IRS; DLR, courtesy of DLR, Oberpfaffenhofen, Germany.)

Floods

Floods can easily be monitored with radar images. Figures 2.117 and 2.118 show multitemporal images of a river flood on the Oder River along the German–Polish boundary.

Forest and Grass Fires

In fire monitoring, three stages are important:

1. Determination of fire hazards—An index with data indicating humidity, wind speed, cloud cover, ground temperature, and green composition of the land can be formed to judge fire potential.
2. After a fire has broken out, NOAA-AVHRR can monitor the extent of the fire in the thermal band. A band 1, 2, and 4 combination can

Figure 2.116   NDVI from DaedalusScanners for an agricultural area. (NDVI image: SI/Antrix/Euromap 1997, GAF 1999, courtesy of GAF Remote Sensing and Information Systems, Munich, Germany.)

Figure 2.117   Radarsat flood image of the Oder River. (© Radarsat 1997, GAF 1997, courtesy of GAF Remote Sensing and Information Systems, Munich, Germany.)

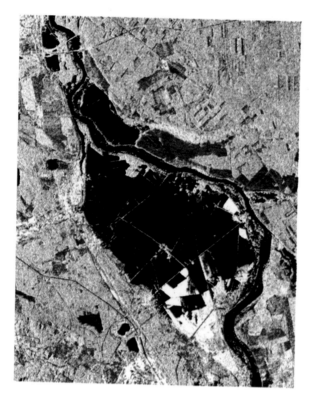

Figure 2.118 Radarsat sequential flood image of the Oder River. (© Radarsat 1997, GAF 1997, courtesy of GAF Remote Sensing and Information Systems, Munich, Germany.)

distinguish smoke and differentiate between burned and unburned areas.

3. After a fire, damage assessment can be made with the help of the images. Figure 2.119 shows the NOAA image of a forest fire along the Siberian–Chinese border.

Environmental awareness has prompted the European Space Agency to launch a series of dedicated satellites for the Global Monitoring for Environment and Security (GMES) program starting with the launch in 2013. This program was renamed in 2013, as "Copernicus."

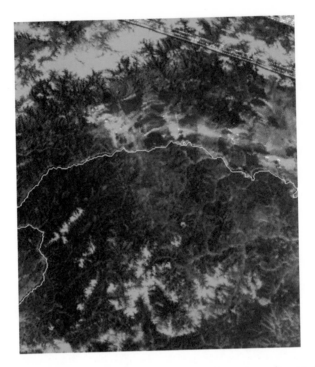

Figure 2.119 Forest fire at the Siberian–Chinese boundary. (Image from CEOS CDROM 98, courtesy of CNES, Toulouse, France.)

The Sentinel satellites focus on:

- Sentinel 1—Synthetic aperture radar applications in C-band to monitor sea ice, the marine environment, and land surface risks
- Sentinel 2—A multispectral earth observation system for disaster relief of the SPOT type
- Sentinel 3—An ocean-oriented satellite to monitor ocean color, sea surface topography, and sea temperature

Chapter 3

Photogrammetry

Photogrammetry is concerned with the geometric measurement of objects in analogue or digital images.

EVOLUTION OF PHOTOGRAMMETRY

The use of photogrammetry is based on the possibility of optically projecting the terrain onto a flat surface, which thus recovers the image by means of a photographic emulsion or by digital sensors. After the invention of photography by Nièpce (Figure 3.1) and Daguerre (Figure 3.2) in 1839, the French military topographer Aimé Laussedat (Figure 3.3) constructed a first camera in 1851, which permitted making measurements on photographs. In 1858, the German architect A. Meydenbauer (Figure 3.4) introduced measurements on photos for the documentation of public buildings.

Single-Image Photogrammetry

Because of the lack of efficient computing facilities, the reconstruction of objects was graphic, following the laws of the perspective that had been developed during the Italian Renaissance in the 15th century (see Figures 3.5 and 3.6 of Leonardo Da Vinci, and Figures 3.7 and 3.8 of Albrecht Dürer).

With the exception of photographs from balloons, taken for military interpretation purposes in the battle of Solferino in 1859 and during the American Civil War, the standard application was terrestrial due to the lack of a suitable aerial photographic platform.

The simplest form of restitution of single images was by rectification. Rectification could be applied to plane surfaces.

Figure 3.1 Joseph Nicephor Niépce, 1765–1833, France. (From the ISPRS Archives; previously published by ISPRS for the 100th Anniversary 2010 by G. Konecny.)

Figure 3.2 Louis Daguerre, 1787–1851, France. (From the ISPRS Archives; previously published by ISPRS for the 100th Anniversary 2010 by G. Konecny.)

Figure 3.3   Aimé Laussedat, 1819–1907, France. (From the ISPRS Archives; previously published by ISPRS for the 100th Anniversary 2010 by G. Konecny.)

Figure 3.4   Albrecht Meydenbauer, 1834–1921, Germany. (From the ISPRS Archives; previously published by ISPRS for the 100th Anniversary 2010 by G. Konecny.)

Figure 3.5   Leonardo da Vinci, 1452–1519, Italy. (From the DGPF German Society for Photogrammetry and Remote Sensing Archives; previously published by DGPF for the 100th Anniversary 2009 by J. Albertz.)

The general projective relations between the coordinates in two planes in their arbitrary coordinate systems $x'y'$ (image plane) and x, y (object plane) are:

$$x_i = \frac{a_1 x_i' + a_2 y_i' + a_3}{a_7 x_i' + a_8 y_i' + 1}$$

$$y_i = \frac{a_4 x_i' + a_5 y_i' + a_6}{a_7 x_i' + a_8 y_i' + 1}$$

This means that the coordinates of four arbitrary points in image and object define the coefficients a_1 to a_8. Based on this graphical or numerical rectification, procedures can be developed for plane surfaces (flat terrain, house walls). Figure 3.9 shows the residual differences, $\Delta r'_E$, due to topography after rectification.

Figure 3.6   Use of perspective by Leonardo da Vinci. (From the ISPRS Archives; previously published by ISPRS for the 100th Anniversary 2010 by G. Konecny.)

Figure 3.7   Albrecht Dürer, 1471–1528, Germany. (From the ISPRS Archives; previously published by ISPRS for the 100th Anniversary 2010 by G. Konecny.)

Figure 3.8 Dürer's book on instructions for use of the perspective. (From the ISPRS Archives; previously published by ISPRS for the 100th Anniversary 2010 by G. Konecny.)

The graphical restitution was extensively used in the 19th century. In the 1920s, special optical rectifiers were built. These had to satisfy two optical conditions. First, the lens equation

$$\frac{1}{a} + \frac{1}{b} = \frac{1}{f}$$

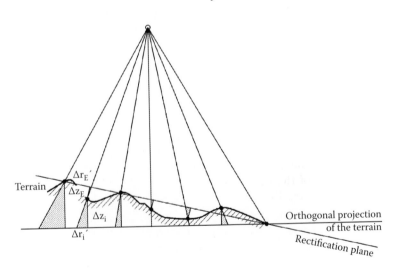

Figure 3.9 Errors in rectification.

with a being the image distance and b the projection distance of the image rectified and f the focal length of the optics of the rectifier.

Second, the so-called Scheimpflug condition, which stated that sharpness could only be reached if the plane of the image, the plane of the objective, and the plane of the projection would intersect along a straight line in space.

Special rectification devices fulfilling these conditions automatically, using mechanical devices, were, however, not built until the mid-1920s.

These also included additional advantages: If the two arbitrary coordinate systems in the image plane and in the projection plane could be related to each other in such a way that their origins were determined by the projection center, then the general projective equations between the two planes could be simplified to:

$$x_i = \frac{a_1 x_i' + a_2 y_i' + 1}{a_5 x_i' + a_6 y_i' + 1}$$

$$y_i = \frac{a_3 x_i' + a_4 y_i' + 1}{a_5 x_i' + a_6 y_i' + 1}$$

Then the rules of perspective geometry can be applied. This permits the execution of a rectification with the coordinates of only three points known in image and object. The mechanical rectifiers required the centering of the images by fiducial marks. Furthermore, a shift of the image plane, d, depending on the focal length of the objective of the rectifier and the focal length of the camera taking the photograph was required

$$d = \left(\frac{f_{image}^2}{f_{rectifier}^2} - 1 + \frac{a^2}{b^2} \right) \cdot \frac{f_{rectifier}}{2} \cdot tg\, v$$

with v being the angle between the objective and projection plane.

The procedure of three-dimensional restitution consisted of measuring image coordinates. When relating these to the projection center identified by fiducial marks of the camera, horizontal and vertical angles to identifiable objects could be derived graphically or numerically. With the terrestrial coordinates of the exposure station determined by ground surveys, and with the exposure directions measured and set in phototheodolites, the coordinates of the object points could be found. Sebastian Finsterwalder used this method of phototopography for the survey of a glacier in 1889 (Figures 3.10 and 3.11). Photogrammetry in this simple form was an added tool to ground survey procedures in inaccessible areas (see Figure 3.12).

Figure 3.10   Sebastian Finsterwalder, 1862–1951, Germany. (From the ISPRS Archives; previously published by ISPRS for the 100th Anniversary 2010 by G. Konecny.)

Analogue Stereo Photogrammetry

It was the stereoscopic measurement principle, developed around 1900, that permitted the automation of the restitution process. The stereocomparator of Carl Pulfrich (Figures 3.13 and 3.14) in Germany (1901) and of Fourcade (see Figure 3.15) in South Africa (1901) permitted the deduction of spatial information through the measurement of observed image parallaxes (see Figure 3.16).

In order to assist stereo viewing and stereo measurement, the terrestrial photographic exposures had to be forced into a rigid configuration, preferring the normal case (see Figure 3.17).

In 1907, Eduard van Orel (Figure 3.18) developed a mechanical plotting device for the reconstruction of rays measured by stereoscopic principles, the Zeiss–Orel Stereoautograph (Figure 3.19) used since 1909 in the survey of Alpine regions. But the more general use of photogrammetry was introduced because of the availability of controllable aerial platforms. The motorized airplane of the Wright brothers (from 1903) soon became an accepted photogrammetric platform, which was used extensively for interpretive purposes during World War I.

In 1915, Oskar Messter (Figure 3.20) developed the aerial survey camera, which permitted a systematic survey of the terrain by near vertical aerial photographs. Even though Sebastian Finsterwalder had already in 1903 found a

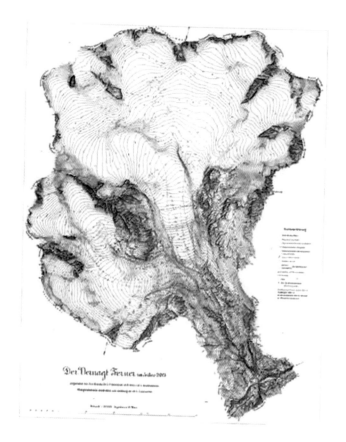

Figure 3.11 Terrestrial photogrammetric survey of the Vernagt Glacier, Tyrol, Austria, 1888–1889. Compiled by hand calculations in 1 year. (From the ISPRS Archives; previously published by ISPRS for the 100th Anniversary 2010 by G. Konecny.)

mathematical restitution procedure of two images taken from balloons, the lack of computing aids prevented use of an analytical reconstruction of the images (Figures 3.21 to 3.23).

Also in 1915, Max Gasser (Figure 3.24) patented and constructed an optical stereoprojection device that could orient two images relative to each other and which could orient both with respect to the terrain (Figure 3.25). The modified instrument was introduced into the market by the Carl Zeiss Company in 1933, under the name of "Multiplex." After the end of World War I, the optical industry in Europe began to construct optical and mechanical devices for the stereo reconstruction of overlapping images.

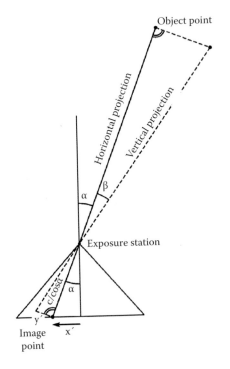

Figure 3.12   Horizontal and vertical angles derived from photocoordinates.

Figure 3.13   Carl Pulfrich, 1858–1927, Germany. (From the ISPRS Archives; previously published by ISPRS for the 100th Anniversary 2010 by G. Konecny.)

Figure 3.14   Pulfrich Stereocomparator built by Carl Zeiss, 1901. (From the ISPRS Archives; previously published by ISPRS for the 100th Anniversary 2010 by G. Konecny.)

Figure 3.15   Henry Georges Fourcade, 1865–1948, South Africa. (From the ISPRS Archives; previously published by ISPRS for the 100th Anniversary 2010 by G. Konecny.)

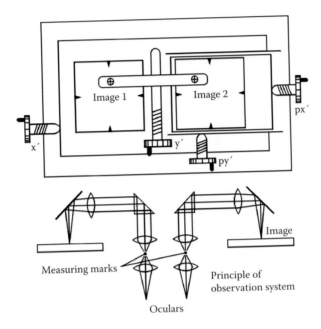

Figure 3.16   Pulfrich stereo comparator principle.

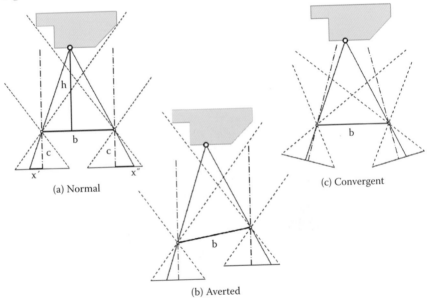

Figure 3.17   (a) Normal, (b) averted, and (c) convergent survey configurations of terrestrial photogrammetry.

Figure 3.18   Eduard von Orel, 1877–1941, Austria. (From the DGPF Archives; previously published by DGPFS for the 100th Anniversary 2009 by J. Albertz.)

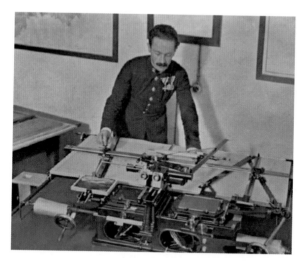

Figure 3.19   Von Orel Stereoautograph, built as a prototype in 1907, manufactured by Carl Zeiss in 1909, for terrestrial photogrammetry. (From the ISPRS Archives; previously published by ISPRS for the 100th Anniversary 2010 by G. Konecny and J. Albertz.)

Figure 3.20 Oskar Messter, 1866–1943, German inventor of the aerial photogrammetric camera, 1915. (From the DGPF Archives; previously published by DGPF for the 100th Anniversary 2009 by J. Albertz.)

Figure 3.21 Balloon image 1 taken by Sebastian Finsterwalder of Gars am Inn, Germany, in 1899. (From the ISPRS Archives; previously published by ISPRS for the 100th Anniversary 2010 by G. Konecny.)

Figure 3.22 Balloon image 2 taken by Sebastian Finsterwalder of Gars am Inn, Germany, in 1899. (From the ISPRS Archives; previously published by ISPRS for the 100th Anniversary 2010 by G. Konecny.)

Initiators of this development were R. Hugershoff (Figure 3.26) in Dresden, Germany (1919), Bauersfeld (Figure 3.27) of Zeiss in Jena, Germany (1921), E. Santoni (Figure 3.28) and U. Nistri (Figure 3.29) in Italy (1921), G. Poivilliers (Figure 3.30) in France (1923), and H. Wild (Figure 3.31) in Switzerland (1926), with E.H. Thompson (Figure 3.32) in England and Bausch & Lomb in the United States to follow.

Measuring Marks

The reconstruction of the geometry of aerial photographs was by optical means or by mechanical means, or a combination of both. The stereo measurement was made possible by an identical measuring mark inserted in the optical path or by a light mark on a projection table (see Figure 3.33). Typical examples of this instrumental development are shown in Figures 3.34 and 3.35.

All analogue stereo restitution instruments are attempting to measure object coordinates x, y, z corresponding to a geometry of vertical photographs (Figure 3.36).

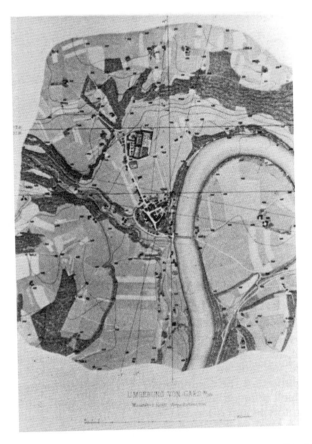

Figure 3.23   Map of Gars am Inn compiled by Sebastian Finsterwalder by analytical calculations by hand, taking 3 years. (From the ISPRS Archives; previously published by ISPRS for the 100th Anniversary 2010 by G. Konecny.)

In vertical photos, the following relations between image coordinates $x'\,y'$ of the first photo and the image coordinates $x''\,y''$ of the second photo separated by the air base, b, are valid:

$$x = -\frac{h}{f}x'$$

$$y = -\frac{h}{f}y'$$

Figure 3.24   Max Gasser, 1872–1954, inventor of the multiplex, 1915, Germany. (From the ISPRS Archives; previously published by ISPRS for the 100th Anniversary 2010 by G. Konecny.)

and

$$h = -\frac{bf}{px} = -\frac{b \cdot f}{(x' - x'')}$$

The horizontal parallax, px, is therefore a measure of height differences.

For vertical photos, the vertical parallax $py = (y' - y'')$ is equal to zero in normal case images. Both image points lie in an epipolar plane.

Figure 3.37 shows in the upper part the geometric displacements of a vertical photograph due to height differences, Δh. The lower part shows horizontal displacements due to the inclination of the photo by the angle, v.

Stereo restitution devices permit the changing and the measuring of coordinates x, y, and z (or h) in the "stereo model." But they must also restore the tilt of the photographic camera and the shifts of the exposure station at the time of the exposure. To do this the plotting instruments allow a rotation of their optical or mechanical cameras around the flight axis by the angle, ω, around the vertical axis by the angle, κ, and perpendicular to it by the angle, ϕ. dx_o is a translation along the base, b; dz_o a translation along the vertical and dy_o perpendicular to it.

If a symmetrical grid of nine points in a vertical photograph is changed in orientation by these six elements—dx_o, dy_o, dz_o, $d\phi$, $d\omega$, or $d\kappa$—then the nine points will change in position, as indicated in Figure 3.38.

Figure 3.25 The Gasser Projector, 1915, rebuilt by Carl Zeiss, 1933, in compact form. (From the ISPRS Archives; previously published by ISPRS for the 100th Anniversary 2010 by G. Konecny.)

These effects can be used in the orientation procedures of aerial photographic diapositives in the stereo plotting instrument. In a stereo model consisting of two aerial photos overlapping by about 60%, six major locations relevant for an orientation can be defined, as shown in Figure 3.39.

Stereo restitution in the plotting instrument can then be carried out in the following steps.

Interior orientation consists of centering the aerial photographic diapositives in the plate holders by help of the fiducial marks and by setting the principal

Figure 3.26   Carl Reinhard von Hugershoff, 1882–1941, built first mechanical stereo-plotter for aerial photos, 1919, Germany. (From the ISPRS Archives; previously published by ISPRS for the 100th Anniversary 2010 by G. Konecny.)

Figure 3.27   Walther Bauersfeld, 1879–1959, designer of the Carl Zeiss Stereoplanigraph, 1921, Germany. (From the ISPRS Archives; previously published by ISPRS for the 100th Anniversary 2010 by G. Konecny.)

Figure 3.28 Ermingeldo Santoni, 1896–1940, designer of the Galileo Stereocartograph in Italy before 1938. (From the ISPRS Archives; previously published by ISPRS for the 100th Anniversary 2010 by G. Konecny.)

Figure 3.29 Umberto Nistri, 1895–1962, founder of Ottico Meccanica Italiana and designer of the photocartograph, 1920, and photogrammetric instruments before 1938. (From the ISPRS Archives; previously published by ISPRS for the 100th Anniversary 2010 by G. Konecny.)

Figure 3.30   Georges Poivilliers, French designer of the Stereotopograph, built by S.O.M. in 1923, France. (From the ISPRS Archives; previously published by ISPRS for the 100th Anniversary 2010 by G. Konecny.)

Figure 3.31   Heinrich Wild, 1877–1951, founder of Wild, Heerbrugg and designer of the Wild Stereoautographs since 1921, Switzerland. (From the ISPRS Archives; previously published by ISPRS for the 100th Anniversary 2010 by G. Konecny.)

Figure 3.32 Edward H. Thompson, designer of Thompson-Watts stereoplotters in Britain, 1952, and analytical photogrammetrist. (From the ISPRS Archives; previously published by ISPRS for the 100th Anniversary 2010 by G. Konecny.)

distance of the instrument to the focal length of the camera. This ensures that a perspective solution can be applied.

Relative orientation of one photo with respect to the other is needed to permit stereo viewing in epipolar planes. Due to the fact that aerial photos, as a rule, are only *nearly* vertical, corresponding points projected from the two photos will not intersect in one point as shown in Figure 3.40, causing horizontal and vertical parallaxes of the corresponding rays. While the horizontal parallax can be removed by height, the vertical parallax is caused by a faulty relative orientation.

The effect of faulty relative orientation on the heights in a stereo model is shown in Figure 3.41.

As early as 1930, Otto von Gruber (Figure 3.42) had introduced an iterative optical–mechanical relative orientation procedure. At the six von Gruber points shown in Figure 3.39, the vertical parallaxes in y are eliminated by the elements shown, while the horizontal parallaxes in x only represent elevation differences between these points. It is also possible to calculate orientation changes for the cameras numerically, if the plotting instrument has dials to change them. The vertical parallaxes in y are measured at the von Gruber points, and the orientation changes are then computed and introduced (Figure 3.43).

In independent pair relative orientation, the rotations of both projectors are used to obtain a y-parallax free stereo model at the von Gruber points. In dependent pair relative orientation, rotations and translations of only one

Identical floating mark for stereo observation

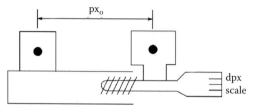

For anaglyphs, alternating shutters and for polarization filters

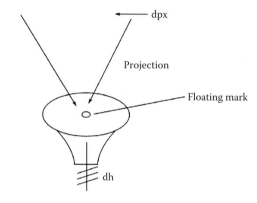

Figure 3.33   Stereo measurement.

projector are used. The functions $f(py_i)$ for dependent pair analogue numerical orientation are, for example

$$bz'' = -\frac{h}{2d}(py_6 - py_4)$$

$$\varphi'' = -\frac{h}{2bd}[(py_5 - py_5) - (py_6 - py_4)]$$

$$\varphi'' = -\frac{h}{4d^2}(2py_1 + 2py_2 - py_3 - py_4 - py_5 - py_6)$$

$$\kappa'' = -\frac{1}{3b}[(py_2 - py_1) + (py_4 - py_3) + (py_6 - py_5)]$$

by'' does not need to be calculated, since it can be visually removed.

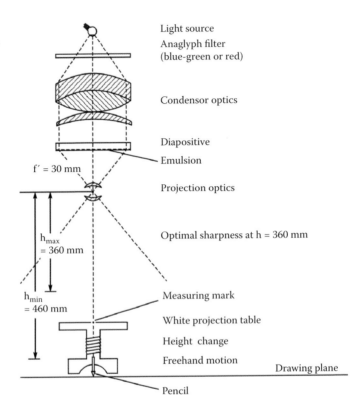

Figure 3.34 Optical stereo restitution instrument multiplex (built by Carl Zeiss, 1933).

Absolute orientation reorients the relatively oriented image pair in space to a ground reference given by control points. One spatial distance between two given points and three elevations spanning a triangle are required for this transformation (see Figure 3.44).

After the orientation has been accomplished, the operator of the plotting instrument can trace the visible details identifiable in the stereo model. Thus, the mapping of streets, rivers and creeks, settlements and houses, and vegetation boundaries is carried out on a map manuscript. Setting the floating mark of the instrument at a certain even height, the contours characterizing the terrain can be drawn.

Plotting instruments even provided the opportunity for control extension along the flight strip by means of analogue aerial triangulation. As long as

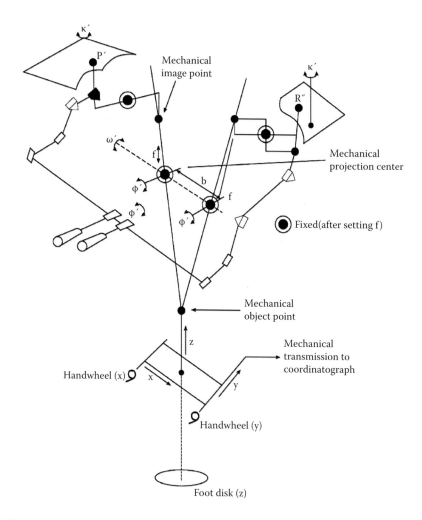

Figure 3.35 Mechanical stereo restitution instrument, Wild A8 (built by Wild, 1950).

identical transfer points in the von Gruber locations were measured in each stereo model, the intersecting rays could be used to reduce the number of required control points in a photogrammetric block.

A great variety of optical and, particularly, mechanical stereo instruments have been produced since World War II by Swiss, German, Italian, French, Russian (see Figure 3.45 for the Russian designer Drobyshev), and British optical manufacturers. The models differed slightly in the ways in which the spatial reconstruction was accomplished. The difference was due mainly to the

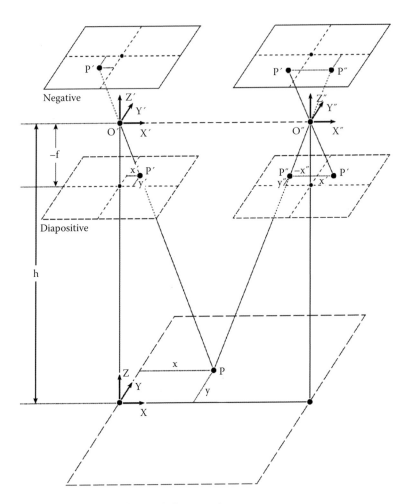

Figure 3.36 Geometry of vertical photographs.

need not to violate patent rights. Nevertheless, the restitution methodology remained common. With these instruments, the current state of topographic mapping in the world has been achieved. This was largely due to the training facilities offered by the ITC Netherlands as part of international cooperation. The founder of the ITC was Willem Schermerhorn (Figure 3.46), a professor at TU Delft and prime minister of the Netherlands from 1945 to 1946.

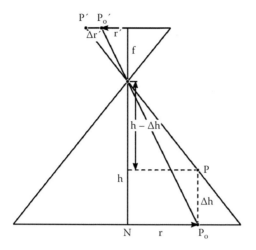

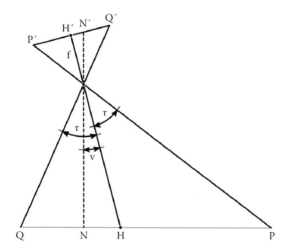

Figure 3.37 Displacements of a photograph due to height differences and tilt.

Calculation of orientation elements κ', φ', κ'', φ'', ω'' or κ'', φ'', ω'', by'', bz'' from parallaxes *pyi*

$$\left.\begin{array}{c} k''\varphi''\omega'' \\ k'\varphi' \end{array}\right\} = f(pyi)$$

$$\left.\begin{array}{c} k''\varphi''\omega'' \\ by''bz'' \end{array}\right\} = f(pyi)$$

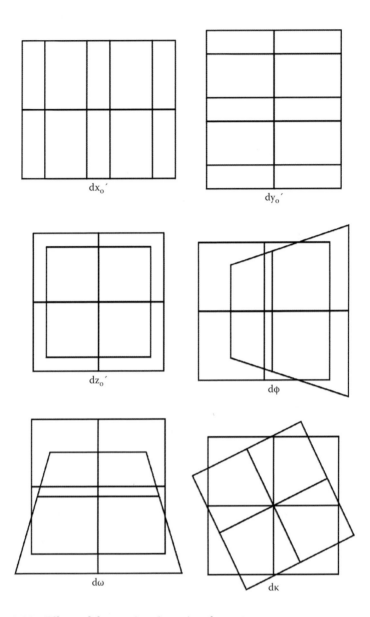

Figure 3.38 Effects of changes in orientation elements.

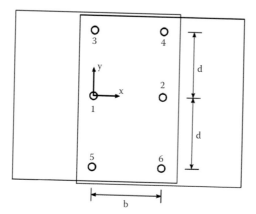

Figure 3.39   Von Gruber points of relative orientation in a stereo model.

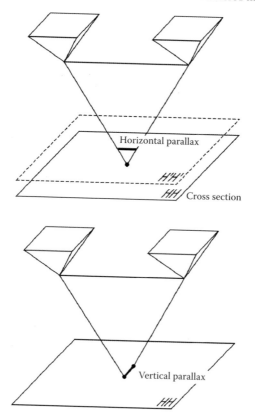

Figure 3.40   Horizontal and vertical parallax.

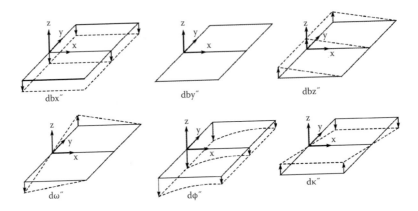

Figure 3.41   Model deformation due to faulty relative orientation.

Figure 3.42   Otto von Gruber. (From the DGPF Archives; previously published by DGPF for the 100th Anniversary 2009 by J. Albertz.)

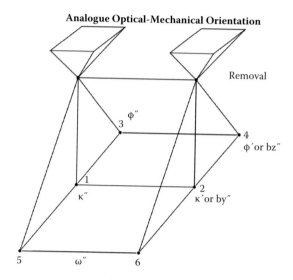

Analogue Optical-Mechanical Orientation

Removal

ϕ''

3

4

ϕ' or bz''

1

κ''

2

κ' or by''

5

ω''

6

Analogue Numerical Orientation

Calculation

py3

py4

$$\left.\begin{array}{l} \kappa''\ \phi''\ \omega'' \\ \kappa'\ \phi' \end{array}\right\} = f(pyi)$$

py1

py2

$$\left.\begin{array}{l} \kappa''\ \phi''\ \omega'' \\ by''\ bz'' \end{array}\right\} = f(pyi)$$

py5

py6

Figure 3.43 Relative orientation.

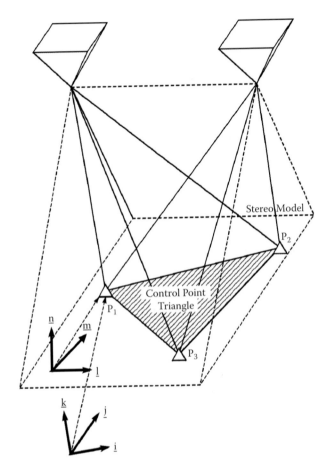

Figure 3.44   Absolute orientation.

In the Western hemisphere and other developed countries, these analogue instruments have now been replaced by analytical and digital restitution instruments. But they continue to be used in developing countries.

Analytical Photogrammetry

Starting in the 1950s, computer systems became available that permitted a rigorous treatment of the photogrammetric solution. This treatment, first applied to point measurement in images or the stereo model by a human operator, not only automated many steps of the photogrammetric restitution process but

Figure 3.45 F.V. Drobyshev, Russian designer of photogrammetric instruments. (From the ISPRS Archives; previously published by ISPRS for the 100th Anniversary 2010 by G. Konecny.)

Figure 3.46 Willem Schermerhorn, 1894–1977, founder of the ITC and prime minister of the Netherlands, 1945–1946. (From the ISPRS Archives; previously published by ISPRS for the 100th Anniversary 2010 by G. Konecny.)

also made photogrammetry more accurate and more reliable by use of least square adjustment and statistical testing included in the computer solutions.

Analytical photogrammetry concerns itself with the modeling of sensor geometry and its restitution. As such, it maintains its full validity even at a time when most restitution operations have turned to the use of digital images and digital image processing. The principles of analytical and digital photogrammetry are discussed later in this chapter.

The advancements brought about by analytical photogrammetry can be summarized thus:

- The measurement of image coordinates of corresponding points in a photographic model or block permitted the calculation and adjustment of the bundle of rays determining the three-dimensional geometry. This solved the need for control extension in photogrammetric blocks. Helmut Schmid (Figure 3.47), Duane Brown (Figure 3.48), Karl Rinner (Figure 3.49), and E.H. Thompson developed the theory in modern algebra, after Earl Church in the United States started to deal with analytical orientation by hand calculations in the 1940s.
- The possibility to rapidly perform digital conversion of coordinates opened the way for a new design of stereo restitution instruments in the form of analytical plotters. Uki Helava (Figure 3.50) invented the analytical plotter principle in 1955. The first analytical plotters were

Figure 3.47 Helmut Schmid, 1914–1998, analytical photogrammetrist, Germany, United States, Switzerland. (From the ISPRS Archives; previously published by ISPRS for the 100th Anniversary 2010 by G. Konecny.)

Figure 3.48   Duane Brown, analytical photogrammetrist, United States. (From the ISPRS Archives; previously published by ISPRS for the 100th Anniversary 2010 by G. Konecny.)

Figure 3.49   Karl Rinner, 1912–1991, theory of analytical photogrammetry Austria. (From the ISPRS Archives; previously published by ISPRS for the 100th Anniversary 2010 by G. Konecny.)

Figure 3.50 Uki V. Helava, inventor of the analytical plotter, Finland, Canada, United States. (From the ISPRS Archives; previously published by ISPRS for the 100th Anniversary 2010 by G. Konecny.)

built by Ottico Meccanica Italiana in cooperation with Bendix USA to satisfy U.S. military requirements to evaluate unconventional imagery (panoramic photography).

The principle of the functioning of an analytical plotter is shown in Figure 3.51.

In analogue stereo plotting devices, model coordinates x, y, z were physically located by the plotter operator in the scale of the stereo model. The observation of corresponding image points was realized by an analogue projection with optical and mechanical devices responsible for the coordinate conversion into the image systems of $x'\,y'$ and $x''\,y''$ coordinates.

An analytical plotter functions in a similar way, except that the real-time coordinate conversion for observation of image points is done on the basis of computation. Servomotors are used to shift the observation system to the computed positions. This has the advantage that the range of computations can be extended to cover geometries, which cannot be handled by the relatively simple optical and mechanical devices.

This means that the geometry of aerial photographs can be corrected on-line for lens and film distortions known from an off-line or on-line calibration process. It means further that stereo restitution can be extended into the use of unconventional imaging geometries provided by optical and microwave imaging devices.

Due to the possibility of steering the analytical plotter automatically to known xyz positions stored in the computer, this offers an automated facility for performing a semiautomatic interior, relative and absolute orientation, for semiautomatic or automatic point transfer from model to model and for the measurement of predetermined elevation grids. This grid determination of

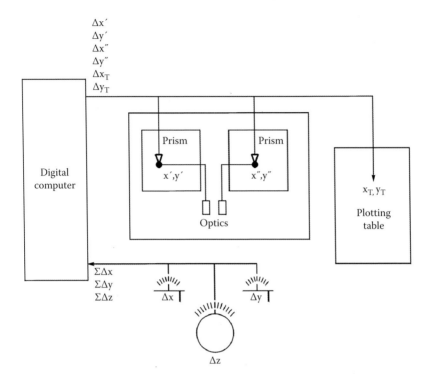

Figure 3.51 Functioning of an analytical plotter.

heights in the analytical plotter performs more rapidly than the tracing of contours, which is standard with analogue instruments.

All measured point coordinates are recorded in the computer files together with an appropriate code. The digital data sets acquired in analytical plotters can therefore be easily incorporated in further GIS processing operations. Analytical plotters have so far successfully competed with digital techniques for the extraction of linear features in topographic mapping, since their stereoviewing and stereo-interpretation capabilities of the film diapositives have generally been superior to the evaluation of digital images in stereo workstations. Both LH Systems and Z/I Imaging marketed analytical plotters until 1990 (Figures 3.52 and 3.53).

The automation capabilities of analytical plotters were also successfully incorporated into the process of differential rectification, in which the image geometry, distorted due to perspective displacements of different terrain heights, was converted into orthoprojected orthophotos.

Even though orthophoto instrumentation began its development in the early 1930s, with O. Lacmann in Germany (Figure 3.54) and R. Ferber in France, and

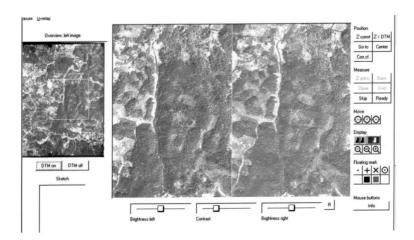

Figure 3.52   The Z/I Imaging Planicomp P3.

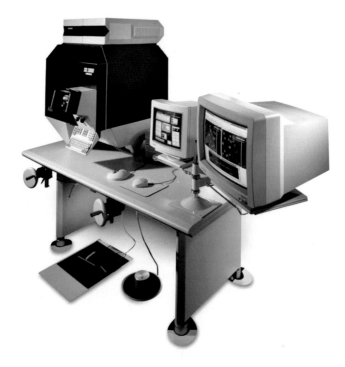

Figure 3.53   The LH Systems SD 2000/3000.

Figure 3.54 Otto Lacmann, 1887–1961, inventor of orthophotography, Germany. (From the DGPF Archives; previously published by DGPF for the 100th Anniversary 2009 by J. Albertz.)

even though the U.S. Geological Survey instituted an orthophoto mapping program in the 1950s with specially constructed analogue devices, orthophotography only became a generally accepted tool after commercial development by Zeiss in Germany, Wild in Switzerland, and OMI in Italy manufactured orthophoto devices on the analytical plotter principle. Figure 3.55 shows the functioning of an analytical orthophoto device.

Analytical orthophoto printers allowed the projection of an image slit onto a drum containing photographic film. While the system permitted the exposure along the drum by moving an optical system along it, the next slits could be exposed by a stepwise shift on the drum.

The analytical plotter controlled the optical projection of the slit, changing its location on the photo, $x'y'$, its magnification, dm, and its rotation, Θ', under computer control during the scan. Analytical orthoprinters have now been replaced by digital orthorectification.

Digital Photogrammetry

Digital photogrammetry makes use of digital or digitized images. This permits vastly extended automation possibilities. Even though digital photogrammetry was already experimentally realized by John Sharp of IBM in

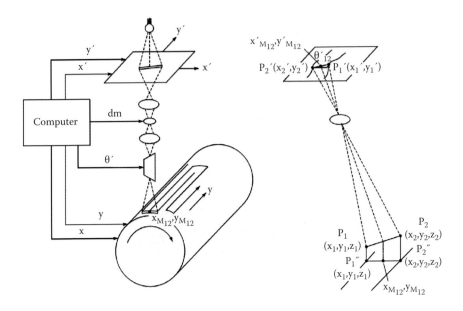

Figure 3.55   Analytical orthophoto device.

1965, it required the development of fast computers with adequate storage facilities before it could become practical. The time arrived around 1988 when the first digital photogrammetric workstations were shown at the ISPRS Kyoto Congress.

In addition to the capabilities offered by analytical plotters, digital workstations now permit the use of newly developed automated technology such as:

- Automated or automatic aerial triangulation
- Derivation of digital elevation models by image matching techniques
- Added display and analysis techniques
- Integration into GIS systems

A digital stereo workstation consists of the following parts:

- Central processing unit with high performance
- 64-bit operating system
- Large enough memory
- Large enough storage system to store a sufficient number of images
- Graphics system to permit stereo viewing and measurement
- Graphic user interface

Figure 3.56 The LH Systems SOCET-SET. (Image courtesy of LH Systems [Leica Geosystems], San Diego, California. © Leica Geosystems, 2002.)

The prime manufacturers of digital workstations have been around for about 10 years.

- LH-Systems with SOCET-SET (Figure 3.56)
- Z/I Imaging with Image Station SSK and with Image Station 2001 or Z4 (Figure 3.57)
- Virtuozo

PRINCIPLES OF ANALYTICAL AND DIGITAL PHOTOGRAMMETRY

Fundamental to the modern treatment of photogrammetry are the tools of matrix algebra and of least square adjustment. With these tools, the problems of spatial networks and their coordinate system conversions can be efficiently treated.

Coordinate Transformations between Image and Terrain

Basic to photogrammetric restitution is the conversion of two-dimensional image coordinates into three-dimensional object coordinates and vice versa. The basic relations are shown in Figure 3.58.

Figure 3.57   The Z/I Imaging Image Station 2001. (Image courtesy of Z/I Imaging Corporation, Oberkochen, Germany.)

Image Coordinates and Local Cartesian Object Coordinates

The relations can be expressed as three-dimensional vectors between the following points:

- O, the origin of the Cartesian object coordinate system, x, y, z
- P_i, the object point with its coordinates, x_i, y_i, z_i, in that system, expressed by the vector $\vec{x}_i$
- O′, the exposure station with its coordinates x_o', y_o', z_o', in that system, expressed by the vector $\vec{x}_o'$
- With these two vectors the vector OP_i forms a spatial triangle

O′P_i can be expressed by the image coordinates, x', y', and f or the vector $\vec{P}_i'$. A prerequisite for the perspective transformation is that the origin of the image coordinate system is linked to the projection center of the exposure station, O′.

Furthermore, the image coordinate system requires a spatial rotation expressed by a three-dimensional rotation matrix, R, and a scale change, λ_{i}', between image measurements and the object coordinates.

The vectors in the triangle OO′P_i can then be added:

$$\vec{x}_i = \vec{x}_o + \lambda_i' R \cdot \vec{p}_i'$$

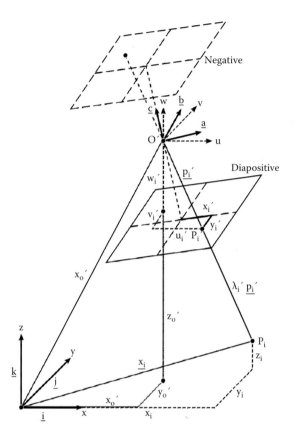

Figure 3.58 Image and object coordinate systems.

Its coordinate components are:

$$\begin{pmatrix} x_i \\ y_i \\ z_i \end{pmatrix} = \begin{pmatrix} x_o' \\ y_o' \\ z_o' \end{pmatrix} + \lambda_i' R \begin{pmatrix} x_i' \\ y_i' \\ -f \end{pmatrix}$$

For the convenience of having the axes of the object and image coordinate systems pointing in similar directions, the principal distance, f, is conventionally entered with a minus sign. Both systems are considered as orthogonal, with their unit vectors $\vec{i}, \vec{j}, \vec{k}$ and $\vec{a}, \vec{b}, \vec{c}$ perpendicular to each other, and each axis has the same scale. In this case the inverse relation does not require calculation of an inverse matrix, R^{-1} from R, but it can be expressed as its transpose, R^T.

$$\begin{pmatrix} x' \\ y' \\ -f \end{pmatrix} = \frac{1}{\lambda_i'} R^T \begin{pmatrix} x_i - x_o' \\ y_i - y_o' \\ z_i - z_o' \end{pmatrix}$$

There are various ways to define the rotational matrix, R. This can be done with the use of direction cosines of the spatial angles between the axes $x'x, x'y, ..., z'z$:

$$R = \begin{pmatrix} \cos(x'x)\cos(y'y)\cos(z'x) \\ \cos(x'y)\cos(y'y)\cos(z'y) \\ \cos(x'z)\cos(y'z)\cos(z'z) \end{pmatrix}$$

It is, however, more convenient to follow the tradition of analogue photogrammetry. In analogue instruments, the rotation around the x-axis was called ω, around the y-axis, φ, and around the z-axis, κ. The rotations R_ω, R_φ, and R_κ could be formed accordingly. In plotting devices, however, the inverse relations converting object to image coordinates were preferred with the rotations R_ω^T, R_φ^T, R_k^T.

A derivation of the coefficients of the rotation matrix R_k^T is shown in Figure 3.59.

$$x' = x \cos \kappa + y \sin \kappa$$
$$y' = y \cos \kappa + x \sin \kappa$$

Thus, the three-dimensional matrix becomes:

$$\begin{pmatrix} x' \\ y' \\ z' \end{pmatrix} = \begin{pmatrix} \cos\kappa & \sin\kappa & 0 \\ -\sin\kappa & \cos\kappa & 0 \\ 0 & 0 & 0 \end{pmatrix} \begin{pmatrix} x \\ y \\ z \end{pmatrix} = R_k^T \begin{pmatrix} x \\ y \\ z \end{pmatrix}$$

$$R_\varphi^T = \begin{pmatrix} \cos\varphi & 0 & \sin\varphi \\ 0 & 1 & 0 \\ -\sin\varphi & 0 & \cos\varphi \end{pmatrix}, \text{ and}$$

$$R_\omega^T = \begin{pmatrix} 1 & 0 & 0 \\ 0 & \cos\omega & \sin\omega \\ 0 & -\sin\omega & \cos\omega \end{pmatrix}$$

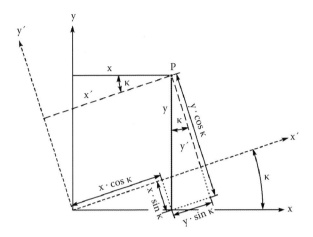

Figure 3.59   Rotational matrix derivation.

Plotting instruments that converted object coordinates into image coordinates were introduced in a sequential manner such that:

$$
\begin{pmatrix} x' \\ y' \\ z' \end{pmatrix} = R_\kappa^T \cdot R_\phi^T \cdot R_\omega^T \begin{pmatrix} x \\ y \\ z \end{pmatrix}
$$

The matrix multiplication of these three sequential matrices results in:

$$
R^T = \begin{pmatrix} r_{11} & r_{21} & r_{31} \\ r_{12} & r_{22} & r_{32} \\ r_{13} & r_{23} & r_{33} \end{pmatrix} = \begin{pmatrix} \cos\kappa\cos\phi & \begin{matrix} \sin\kappa\cos\omega \\ -\cos\kappa\sin\phi\sin\omega \end{matrix} & \begin{matrix} \sin\kappa\cos\omega \\ +\cos\kappa\sin\phi\cos\omega \end{matrix} \\ -\sin\kappa\cos\phi & \begin{matrix} \cos\kappa\cos\omega \\ +\sin\kappa\sin\phi\sin\omega \end{matrix} & \begin{matrix} \cos\kappa\sin\omega \\ -\sin\kappa\sin\phi\sin\omega \end{matrix} \\ -\sin\phi & -\cos\phi\sin\omega & \cos\phi\cos\omega \end{pmatrix}
$$

Since these coefficients must all represent the direction cosines of the angles between the axes, it is possible to derive the matrix, R^T, also with other sign conventions and other rotational sequences. The resultant coefficients of R^T can thus be used to calculate the respective ω, ϕ, κ values by the simple comparison of the coefficients.

The object coordinate system x, y, z is a local Cartesian coordinate system that must be related to the geodetic coordinate system based on a reference frame in use in a particular country. The relations between geodetic coordinates are available in geodetic literature.

In a single model or in a small area, the difference between the geodetic reference and the Cartesian object model is so small that it has often been neglected in practice. For a more thorough treatment, an example for converting 3° transverse Mercator (Gauss–Krüger) coordinates into a local Cartesian system is given next.

Transverse Mercator and Geographic Coordinates

Control points are usually known in 3° transverse Mercator coordinates, X_i, Y_i. These can be transformed into geographic coordinates, φ_i, λ_i (see Figure 3.60).

$$\varphi_i = \varphi_F - \frac{\rho}{2N_F^2} t_F (1+\eta_F^2) Y_i^2 + \frac{\rho}{24N_F^4} t_F (5+3t_F^2+6\eta_F^2-6t_F^2-6t_F^2\eta_F^2) Y_i^4$$

with

$$\varphi_F = V_F^2 \left[\frac{X_i}{N_F} - \frac{3}{2}\eta_F^2 t_F \frac{X_i^2}{N_F^2} + \frac{X_i^3}{N_F^3} \cdot \frac{\eta_F^2}{2} \left(t_F -1-\eta_F^2 +5\eta_F^2 t_F \right)^2 \right]$$

$$V_F = \sqrt{1+e'^2 \cos^2 \varphi_F}$$

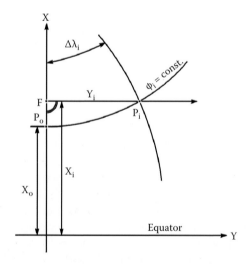

Figure 3.60   Conversion between transverse Mercator and geographic coordinates.

$$N_F = \frac{a}{[(1-e^2)\sin^2\varphi_F]^{1/2}}$$

$$n_F = (e'^2\cos^2\varphi_F)^{1/2}$$

$$e = \sqrt{\frac{a^2-b^2}{a^2}}, e'\sqrt{\frac{a^2-b^2}{b^2}}, p^0 = \frac{180^0}{\pi}$$

$$t_F = t_g\varphi_F$$

a and b are parameters of the chosen reference ellipsoid. λ_F is the chosen principal meridian. φ_F is iteratively calculated.

The inverse transformation is:

$$X_i = X_o + \frac{1}{2}\frac{N_i}{p^2}\cos^2\varphi_i t_i(\lambda-\lambda_o)^2 + \frac{1}{24}\frac{N_i}{p^4}\cos^2\varphi_i t_i(5-t_i^2+9\eta_i^2+4\eta_i^4)(\lambda_i-\lambda_o)^4$$

$$Y_i = \frac{1}{p}N_i\cos\varphi_i(\lambda_i-\lambda_o) + \frac{1}{6}\frac{N_i}{p^3}\cos^3\varphi_i(1-t_i^2+\eta_i^2)(\lambda_i-\lambda_o)^3$$

$$+\frac{1}{120}\frac{N_i}{p^5}\cos^5\varphi_i(5-18t_i^4+14\eta_i^2-58t_i^2\eta_i^2)(\lambda_i-\lambda_o)^5$$

with

$$t_i = t_g\varphi_i$$

$$N_i = \frac{a}{1+e^2\sin^2\varphi_i)^{1/2}}$$

$$\eta_i = (e'^2\sin^2\varphi_i)^{1/2}$$

$$V_i = \sqrt{1+e'^2\cos^2\varphi}$$

$$X_o = N_i\left[\frac{\varphi_i}{V_i^2} + \left(\frac{\varphi_i}{V_i^2}\right)^2 \frac{3}{2}\eta_i^2 t_i - \left(\frac{\varphi}{V_i^2}\right)^3 \frac{\eta_i^2}{2}(t_i^2-1-\eta_i^2-4\eta_i^2-t_i^2)\right]$$

Geographic Coordinates and Geocentric Cartesian Coordinates

Geographic coordinates, φ_i, λ_i, can be converted into geocentric Cartesian coordinates (space rectangular coordinates) X_i', Y_i', Z_i' (see Figure 3.61).

$$X_i' = (N_i + h_i)\cos\varphi_i \cos\lambda_i$$

$$Y_i' = (N_i + h_i)\cos\varphi_i \sin\lambda_i$$

$$Z_i' = \left[N_i(1 = e^2) + h_i \right]\sin\varphi_i$$

The ellipsoidal height, h_i, of a point is composed of the orthometric height, H_i, minus the geoidal undulation, G_i:

$$h_i = H_i - G_i$$

The inverse solution is:

$$tg\lambda_i = \frac{Y_i'}{X_i'}$$

$$(N_i + h_i) = \sqrt{X_i'^2 + Y_i'^2 + (Z_i'^2 + t_1)^2}$$

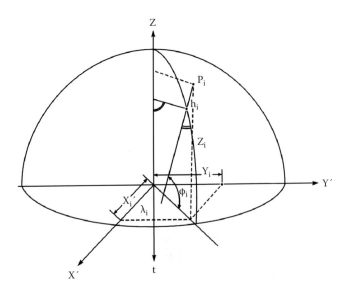

Figure 3.61   Geocentric Cartesian coordinates.

Since $t_1 = e^2 Z'^2$ an iterative solution is required using $(Z_i'^2 + t) = (N_i + h)\sin\varphi_1$ with

$$\varphi_1 = \arcsin\frac{Z_i' + t_1}{(X_i'^2 + Y_i'^2 + (Z_i'^2 + t_1)^2)^{1/2}}$$

with this

$$N_1 = \frac{a}{\sqrt{1 - e^2 \sin\varphi_1}} \quad \text{and } t_2 = N_1 e^2 \sin\varphi$$

After a few iterations, φ_i is obtained as well as:

$$h_i = \sqrt{(X_i'^2 + Y_i'^2 + (Z_i'^2 + t_1)^2} - N_i$$

Transformation between Geocentric and Local Cartesian Coordinates

Transformation of the geocentric Cartesian coordinate system into a local one is possible by the following transformation:

$$\begin{pmatrix} x_i \\ y_i \\ z_i \end{pmatrix} = m \begin{pmatrix} -\sin\lambda_o & \cos\lambda & 0 \\ -\sin\varphi_o\cos\lambda_o & -\sin\varphi_o\sin\lambda_o & \cos\varphi_o \\ \cos\varphi_o\cos\lambda_o & \cos\varphi_o\sin\lambda_o & \sin\varphi_o \end{pmatrix} \begin{pmatrix} x_i' - x_o' \\ y_i' - y_o' \\ z_i' - z_o' \end{pmatrix}$$

$$= mM \begin{pmatrix} X_i' - X_o' \\ Y_i' - Y_o' \\ Z_i' - Z_o' \end{pmatrix}$$

with X_o', Y_o', Z_o', φ_o, and λ_o defining the origin of the local system. A scale factor, m, may also be introduced.

The inverse solution is:

$$\begin{pmatrix} X_i' \\ Y_i' \\ Z_i' \end{pmatrix} = \begin{pmatrix} X_o' \\ Y_o' \\ Z_o' \end{pmatrix} + \frac{1}{m} M^T \begin{pmatrix} x_i \\ y_i \\ z_i \end{pmatrix}$$

Control point coordinates may be converted by these transformations into the photogrammetric object coordinate system, in which the analytical solution is computed. Thereafter, all local coordinates can be retransformed into the coordinates of the projection used.

The reason that this conversion process was neglected in the past was due to the use of a great number of control points in the block between which the transformation errors have been interpolated. The same is true for the apparent height deformation due to earth curvature, d (see Figure 3.62), which amounts to about:

$$d \approx \frac{r^2}{R}$$

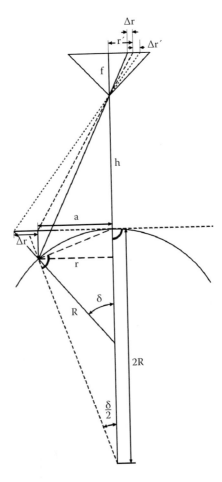

Figure 3.62   Earth curvature.

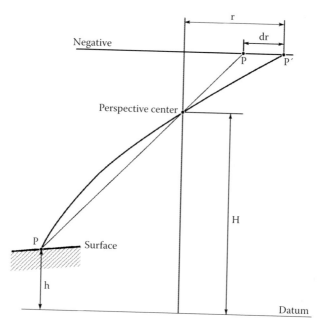

Figure 3.63   Atmospheric refraction.

The same holds true for atmospheric refraction, which acts in the form of a radial distortion (Figure 3.63).

Space Intersection

For the determination of spatial coordinates, at least two photographs from different exposure stations are required. It then becomes possible to apply the transformation between object coordinates x_i, y_i, z_i and photo coordinates of the imaged points in both photos $x_i'y_i'$ and $x_i''y_i''$ (Figure 3.64).

$$\vec{x}_i = \vec{x}_o + \lambda_i' R' \vec{p}_i' = \vec{x}_i = \vec{x}_o'' + \lambda_i'' R'' \vec{p}_i'' \text{ with the components:}$$

$$\begin{pmatrix} x_i' \\ y_i' \\ z_i' \end{pmatrix} = \begin{pmatrix} x_o' \\ y_o' \\ z_o' \end{pmatrix} + \lambda_i' R' \begin{pmatrix} x_i' \\ y_i' \\ -f \end{pmatrix} = \begin{pmatrix} x_o'' \\ y_o'' \\ z_o'' \end{pmatrix} + \lambda''_i R'' \begin{pmatrix} x_i'' \\ y_i'' \\ -f \end{pmatrix}$$

If the coordinates of both exposure stations, $\vec{x}_o'$ and $\vec{x}_o''$, and their photo orientations expressed through their two rotational matrices, R' as a function of

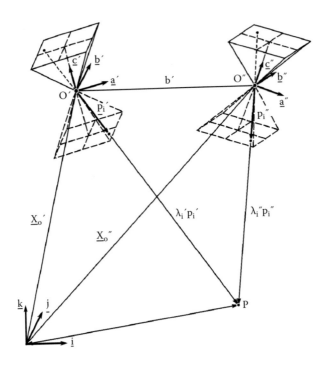

Figure 3.64 Space intersection.

ω', φ', κ', and R'' as a function of ω'', φ'', κ'' are known, then two of the equations may be used to determine the unknown scale factors, λ_i' and λ_i''.

If the projections of the photo coordinate vectors, $\vec{p}_i'$ and $\vec{p}_i'$, in the direction of the object coordinate system are defined as

$$R' \begin{pmatrix} x_i' \\ y_i' \\ -f \end{pmatrix} = \begin{pmatrix} u' \\ v' \\ w' \end{pmatrix}$$

and

$$R'' \begin{pmatrix} x_i'' \\ y_i'' \\ -f \end{pmatrix} = \begin{pmatrix} u'' \\ v'' \\ w'' \end{pmatrix}$$

then the equations can be written as:

$$\begin{pmatrix} x_i \\ y_i \\ z_i \end{pmatrix} = \begin{pmatrix} x_o' \\ y_o' \\ z_0' \end{pmatrix} + \lambda_i' \begin{pmatrix} u_i' \\ v_i' \\ w_i' \end{pmatrix} = \begin{pmatrix} x_o'' \\ y_o'' \\ z_o'' \end{pmatrix} + \lambda_i'' \begin{pmatrix} u_i'' \\ v_i'' \\ w_i'' \end{pmatrix}$$

The two equations with the largest coordinate differences $(x_o'' - x_o') = bx$ and $(y_o'' - y_o') = by$ may be used to solve for λ_i' and λ_i''

$$\lambda_i' = \frac{bx \cdot v_i'' - by \cdot u_i''}{u_i' v_i'' - v' u_i''}$$

$$\lambda_i'' = \frac{by \cdot u_i' - bx \cdot v_i'}{v_i' \cdot u_i'' - u_i' \cdot v_i''}$$

with λ_i' and λ_i'' known, the coordinates of the intersected point, P_i (x_i, y_i, z_i) may be calculated.

Space Resection

Unless the coordinates of the exposure station, x_o', y_o', and z_o', can be directly measured by GPS and unless the components of the rotational matrix, R', as functions of the angles, ω', φ', κ', can be directly surveyed by inertial measuring units (IMUs), the orientation of a single photograph consisting of the six orientation elements, x_o', y_o', z_o', ω', φ', κ', must be determined by a space resection based on known control points identifiable in the photo (see Figure 3.65).

For this, the equations of coordinate conversion between object and photo can be used

$$\vec{x}_i = \vec{x}_o' + \lambda_i' \cdot R \cdot p_i'$$

with its components

$$\begin{pmatrix} x_i \\ y_i \\ z_i \end{pmatrix} = \begin{pmatrix} x_o' \\ y_o' \\ z_o' \end{pmatrix} + \lambda_i' \begin{pmatrix} r_{11} & r_{12} & r_{13} \\ r_{21} & r_{22} & r_{23} \\ r_{31} & r_{32} & r_{33} \end{pmatrix} \begin{pmatrix} x_i' \\ y_i' \\ -f \end{pmatrix}$$

In these equations, the photo coordinates are the observations and the orientation elements, x_o', y_o', z_o', ω', φ, κ, as well as λ_i, the unknown.

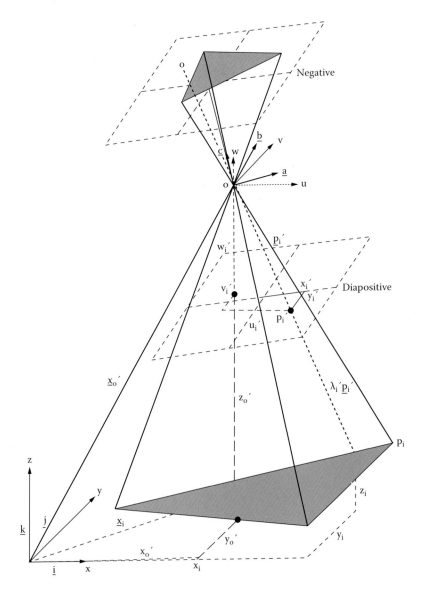

Figure 3.65   Space resection.

It is therefore more convenient to use the inverse relations:

$$
\begin{pmatrix} x_i' \\ y_i' \\ -f \end{pmatrix} = \frac{1}{\lambda_i'} \begin{pmatrix} r_{11} & r_{21} & r_{31} \\ r_{12} & r_{22} & r_{32} \\ r_{13} & r_{23} & r_{33} \end{pmatrix} \begin{pmatrix} x_i - x_o' \\ y_i - y_o' \\ z_i - z_o' \end{pmatrix}
$$

Since f is a constant of the camera used, it is possible to eliminate λ_i' by dividing the first and second equation through the third. This will result in the collinearity equations, which are an expression or a condition, that image point, projection center, and object point must lie on a straight line:

$$
x_i' = -f \frac{r_{11}(x_i - x_o') + r_{21}(y_i - y_o') + r_{31}(z_i - z_o')}{r_{13}(x_i - x_o') + r_{23}(y_i - y_o') + r_{33}(z_i - z_o')}
$$

$$
y_i' = -f \frac{r_{12}(x_i - x_o') + r_{22}(y_i - y_o') + r_{33}(z_i - z_o')}{r_{13}(x_i - x_o') + r_{23}(y_i - y_o') + r_{33}(z_i - z_o')}
$$

With respect to the six unknowns, these are nonlinear equations. To solve them for the required number of points, they are linearized by a Taylor series using its linear terms as a first approximation, with iterations to follow:

$$
x_i' = (x_i')_o + \frac{\delta x_i'}{\delta x_o'} dx_o' + \frac{\delta x_i'}{\delta y_o'} dy_o' + \frac{\delta x_i'}{\delta z_o'} dz_o' + \frac{\delta x_i'}{\delta \omega_o'} d\omega_o' + \frac{\delta x_i'}{\delta \varphi_o'} d\varphi_o' + \frac{\delta x_i'}{\delta \kappa_o'} d\kappa'
$$

$$
y_i' = (y_i')_o + \frac{\delta y_i'}{\delta x_o'} dx_o' + \frac{\delta y_i'}{\delta y_o'} dy_o' + \frac{\delta y_i'}{\delta z_o'} dz_o' + \frac{\delta y_i'}{\delta \omega_o'} d\omega_o' + \frac{\delta y_i'}{\delta \varphi_o'} d\varphi_o' + \frac{\delta y_i'}{\delta \kappa_o'} d\kappa'
$$

The differential coefficients are after differentiation of the collinearity equations as follows:

$$
\frac{\delta x_i'}{\delta x_o'} = -\frac{1}{D}(-x_i' r_{13} - f r_{11})
$$

$$
\frac{\delta x_i'}{\delta y_o'} = -\frac{1}{D}(-x_i' r_{23} - f r_{21})
$$

$$
\frac{\delta x_i'}{\delta z_o'} = -\frac{1}{D}(-x_i' r_{33} - f r_{31})
$$

$$\frac{\delta x_i'}{\delta \omega_o'} = -\frac{1}{D}\left\{ x_i'\left[(y_i - y_o')r_{33} + (z_i - z_o')r_{23}\right] + f\left[-(y_i - y_o')r_{31} + (z_i - z_o')r_{21}\right]\right\}$$

$$\frac{\delta x_i'}{\delta \varphi_o'} = -\frac{1}{D}\left\{ \begin{matrix} x_i'\left[-(x_i - x_o')\cos\varphi' + (y_i - y_o')\sin\varphi'\sin\omega' - (z_i - z_o')\sin\varphi'\cos\omega'\right] \\ + f\left[-(x_i - x_o')\cos\kappa'\sin\varphi' - (y_i - y_o')\cos\kappa'\cos\omega' + (z_i - z_o')\cos\kappa'\cos\varphi'\cos\omega'\right] \end{matrix} \right\}$$

$$\frac{\delta x_i'}{\delta \kappa_o'} = -\frac{1}{D}\left\{ f\left[(x_i - x_o')r_{12} + (y_i - y_o')r_{22} + (z_i - z_o')r_{32}\right]\right\}$$

$$\frac{\delta y_i'}{\delta x_o'} = -\frac{1}{D}(-y_i'r_{13} - fr_{12})$$

$$\frac{\delta y_i'}{\delta y_o'} = -\frac{1}{D}(-y_i'r_{23} - fr_{22})$$

$$\frac{\delta y_i'}{\delta z_o'} = -\frac{1}{D}(-y_i'r_{33} - fr_{32})$$

$$\frac{\delta y_i'}{\delta \omega_o'} = -\frac{1}{D}\left\{ y_i'\left[-(y_i - y_o')r_{33} + (z_i - z_o')r_{23}\right] + f\left[-(y_i - y_o')r_{32} + (z_i - z_o')r_{22}\right]\right\}$$

$$\frac{\delta y_i'}{\delta \varphi_o'} = -\frac{1}{D}\left\{ \begin{matrix} y_i'\left[-(x_i - x_o')\cos\varphi' + (y_i - y_o')\sin\varphi'\sin\omega' - (z_i - z_o')\sin\varphi'\cos\omega'\right] \\ + f\left[-(x_i - x_o')\sin\kappa'\sin\varphi' - (y_i - y_o')\sin\kappa'\cos\omega'\sin\omega' - (z_i - z_o')\sin\varphi'\cos\varphi'\cos\omega'\right] \end{matrix} \right\}$$

$$\frac{\delta y_i'}{\delta \kappa_o'} = -\frac{1}{D}\left\{ f\left[-(x_i - x_o')r_{11} - (y_i - y_o')r_{21} - (z_i - z_o')r_{31}\right]\right\}$$

with the abbreviation

$$D = r_{11}(x_i - x_o') + r_{23}(y_i - y_o') + r_{21}\times(z_i - z_o')$$

The first approximations for $(x_i')_o$ and $(y_i')_o$ are derived from a flight plan and the assumptions of vertical photography:

$$(x_i')_o = -f\frac{x_i - (x_o')_o}{z_i - (z_o')_o}$$

$$(y_i')_o = -f\frac{y_i - (y_o')_o}{z_i - (z_o')_o}$$

In matrix notation, the equation system can be written as:

$$A \overrightarrow{x_{j,1}} = \overrightarrow{\ell_{i,1}}$$
$$_{i,j}$$

in which A_{ij} is the coefficient matrix, $\overrightarrow{x_{j,1}}$ the six orientation unknowns, and $\overrightarrow{\ell_{j,1}}$ the observations i. In order to determine the six orientation parameters, a minimum of three control points are required. They will permit the setting up of six observation equations for these three points:

$$x_i', y_i'$$

These permit solving for the unknowns:

$$x_{6,1} = A_{6,6}^{-1} \ell_{6,1}$$

The inverse of the coefficient matrix can be calculated if $|A| \neq 0$. This is the case if the three control points form a spatial triangle. If more than three control points are available, then the number of observations will be greater than six (e.g., eight observations for four points), and the determination of the orientation parameters will be subject to a least squares adjustment.

For the equations of all point residuals, v_{xi}', v_{yi}' must be introduced, so that

$$v_{i,1} = A_{ij}, x_{j,1} - \ell_{i,1}$$

or in the example

$$v = A \, x - \ell$$
$$_{8,1} \quad _{8,6} {}_{6,1} \quad _{8,1}$$

A least squares solution is obtained if the residuals are made to a minimum. This is the case if:

$$\frac{\delta v_i^T v_i}{\delta x_j} = 0 = A_{j,i}^T A_{i,j} x_{j,i} - A_{j,i}^T \ell_{i,1}$$

Thus

$$x_{j,1} = (A_{j,i}^T A_{i,j})^{-1} A_{j,i}^T \ell_{i,1}$$

or

$$x_{6,1} = (A_{6,8}^T A_{8,6})^{-1} A_{6,8}^T \ell_{8,1}$$

The approximations for x_j can now be replaced by these values and a new iteration can start. The procedure is repeated until no changes in x_j become noticeable.

With the orientation parameters calculated in this manner for all photographs, a space intersection for the determination of object coordinates can follow without the need for a relative and absolute orientation of the stereo model.

Aerial Triangulation

The problem of space resection can be extended to include the simultaneous orientation of multiple exposure stations, which are common in a photogrammetric block. For this, it is necessary to include observations of transfer points, which have been identified and measured in adjacent photos. The object coordinates of these points must be added to the equation system as three additional unknowns per transfer point so that the linearized collinearity equations become:

$$x_i' = (x_i')_o + \frac{\delta x_i'}{\delta x_o'} dx_o' + \frac{\delta x_i'}{\delta y_o'} dy_o' + \frac{\delta x_i'}{\delta z_o'} dz_o' + \frac{\delta x_i'}{\delta \omega'} d\omega' + \frac{\delta x_i'}{\delta \varphi'} d\varphi_o' + \frac{\delta x_i'}{\delta \kappa'} d\kappa'$$

$$+ \frac{x_i'}{\delta x_i} dx_o' + \frac{\delta x_i'}{\delta y_i} dy_o' + \frac{\delta x_i'}{\delta z_i} dz_o'$$

$$y_i' = (y_i')_o + \frac{\delta y_i'}{\delta x_o'} dx_o' + \frac{\delta y_i'}{\delta y_o'} dy_o' + \frac{\delta y_i'}{\delta z_o'} dz_o' + \frac{\delta y_i'}{\delta \omega'} d\omega' + \frac{\delta y_i'}{\delta \varphi'} d\varphi_o' + \frac{\delta y_i'}{\delta \kappa'} d\kappa'$$

$$+ \frac{\delta y_i'}{\delta x_i} dx_i + \frac{\delta y_i'}{\delta y_i} dy_i + \frac{\delta y_i'}{\delta z_i} dz_i$$

These new differential coefficients become, after differentiation of the collinearity equations:

$$\frac{\delta x_i'}{\delta x_i} = -\frac{1}{D}(x_i' \cdot r_{13} + f r_{11})$$

$$\frac{\delta x_i'}{\delta y_i} = -\frac{1}{D}(x_i' \cdot r_{23} + f r_{21})$$

$$\frac{\delta x_i'}{\delta z_i} = -\frac{1}{D}(x_i' \cdot r_{33} + f r_{32})$$

$$\frac{\delta y_i'}{\delta x_i} = -\frac{1}{D}(y_i' \cdot r_{13} + f r_{12})$$

$$\frac{\delta y_i'}{\delta y_i} = -\frac{1}{D}(y_i' \cdot r_{23} - fr_{22})$$

$$\frac{\delta y_i'}{\delta z_i} = -\frac{1}{D}(y_i' \cdot r_{33} + fr_{32})$$

These equations can be applied to all measured points of the block in preparation for a bundle block adjustment. The configuration of photographic bundles and their relationship to stereo models is shown in Figure 3.66.

Figure 3.67 illustrates the application to a block consisting of 4 photos each in 3 flight strips (total of 12 photos) with 28 transfer points in the von Gruber locations. There are $12 \times 6 = 72$ orientation unknowns and there are $28 \times 3 = 84$ transfer point unknowns in the block (total 156 unknowns, x_j). Photos 1, 2, 3, 10, 11, and 12 contain the measurements $x_i'y_i'$ of six transfer points (12 observation equations each) and photos 4, 5, 6, 7, 8, and 9 contain nine transfer points (18 observation equations each). This results in 180 observation equations for this relatively small block

$$v_{180,1} = A_{180,156}x_{156,1} - \ell_{180,1}$$

requiring to invert a matrix of 156×156 equations:

$$x_{156,1} = (A_{156,180}^T \cdot A_{180,156}^T)^{-1}(A_{156,180}^T \cdot \ell_{180,1})$$

The matrix is invertible, if $|A^TA|^{-1} \neq 0$, requiring the knowledge of a minimum number of ground control points. A typical control point configuration in a block is shown in Figure 3.68. The big dots represent horizontal control points and the small dots represent vertical control points.

If some of the transfer points used are ground control points, then their coordinate unknowns can be directly eliminated. To reach the condition, 4 well-placed ground control points are needed. Thus, 12 coordinate unknowns can be eliminated to reach a solution.

It is, however, preferable to leave the equation system untouched and to add the ground control values as additional observations, with the advantage that they can be appropriately weighted with regard to the measurements in the photos. For the 4 ground control points, there will be 12 additional equations.

Each observation equation can be assigned a weight, p_i. It is related to the variance of the observation:

$$p_i = \frac{1}{\sigma_i^2}$$

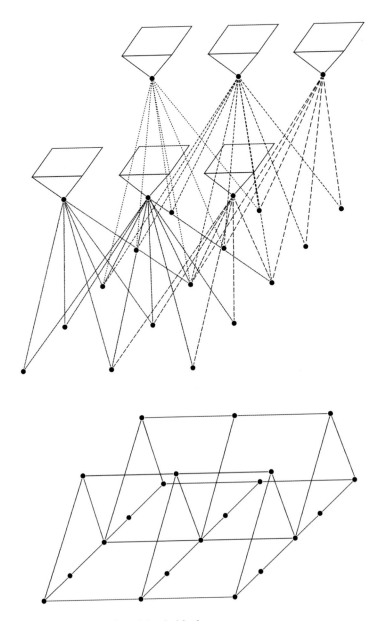

Figure 3.66   Photos and models of a block.

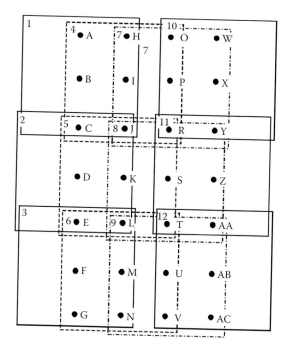

Figure 3.67 Example for a block of 4 × 3 photographs.

Photo coordinate measurements can be assigned a weight commensurate with the standard deviation, σ_p, of measurement (e.g., 8 μm). The ground coordinates of control points can be entered with a weight

$$p_x = \frac{1}{\sigma_x^2}$$

commensurate with their accuracy of point determination.

The least squares adjustment principle will then be

$$v^T p v = (Ax - \ell)^T p(Ax - \ell) = \min$$

with the solution

$$x = (A^T p A)^{-1} A^T p \ell$$

yielding a

$$\sigma_o^2 = \frac{v^T p v}{n - u}$$

with n being the number of observations and u the number of unknowns.

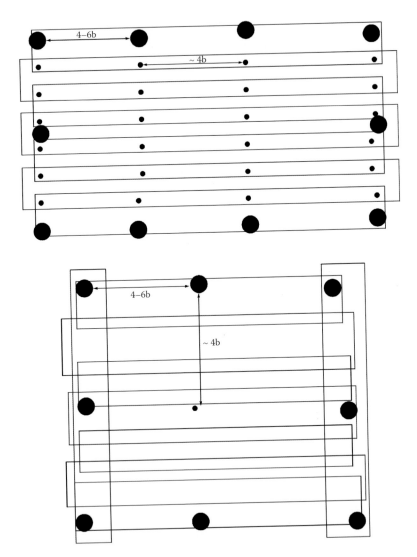

Figure 3.68 Typical control point configuration in a block.

The use of added observations with weights is also possible for direct measurement of exposure stations via GPS and of camera orientations via IMU.

Modern bundle block adjustment programs, such as BLUH established at the University of Hannover, are capable of treating 6000 photo orientations and 200,000 ground points in a simultaneous adjustment.

As the computational power needed for matrix inversion increases with the third power of the size of the matrix, the symmetry of the arrangement of observation helps in matrix partitioning with the possibility to invert partial inverses more efficiently to permit a solution for the 600,000-plus unknowns.

The matrix *A* for the example selected in Figure 3.67 is shown in Figure 3.69.

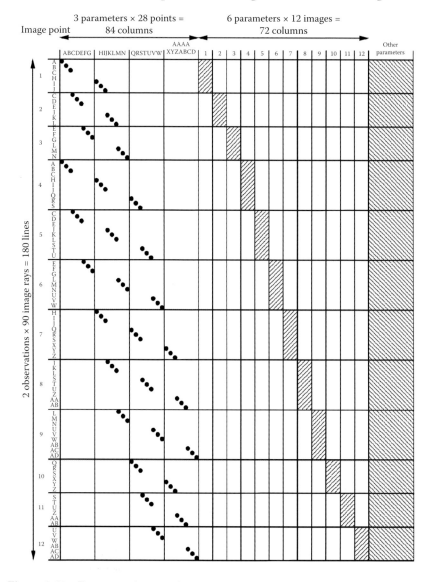

Figure 3.69   Error equation matrix.

The error equations do not show correlations for the orientation elements of different exposure stations; the point coordinates are also uncorrelated. Only a limited number of parameters are correlated. When the normal equations, A^TpA, are formed, the correlations between points and photos become visible (see Figure 3.70).

This permits the grouping of matrix A for the error equations into three parts:

$$v_i' = (A_1 A_2 A_3) \begin{pmatrix} x_1 \\ y_2 \\ z_3 \end{pmatrix} - \ell_i$$

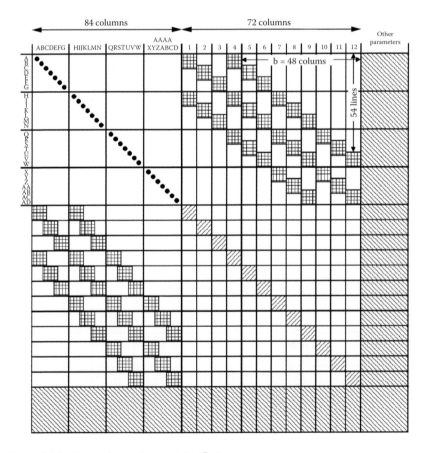

Figure 3.70   Normal equation matrix A^TpA.

A_1 contains the matrix part for the point unknowns, A_2 the matrix part for the orientation unknowns, and A_3 any additional parameters. The normal equation matrix, $A^T pA$, thus is composed of the following groups:

$$\begin{pmatrix} A_1^T A_1 & A_1^T A_2 & A_1^T A_3 \\ A_2^T A_1 & A_2^T A_2 & A_3^T A_3 \\ A_3^T A_1 & A_3^T A_2 & A_3^T A_3 \end{pmatrix} \begin{pmatrix} x_1 \\ x_2 \\ x_3 \end{pmatrix} = \begin{pmatrix} A_1^T \ell \\ A_2^T \ell \\ A_3^T \ell \end{pmatrix}$$

Parts of this matrix do not need to be fully inverted in the solution. $A_2^T A_2$ only consists of 3×3 elements for unknown points, which can easily be inverted. This results in additions to $A_2^T A_2$, which has uncorrelated 6×6 elements. The additions of the partially inverted $A_2^T A_1$ matrix and the matrix parts $A_2^T A_1$ or $A_1^T A_2$ structure matrix $A_2^T A_2$ into a bandwidth structure around the diagonal. The inversion is possible by recursive partitioning. Finally, only $A_3^T A_3$ needs to be fully inverted.

The additional parameters x_3 consist of self-calibration parameters, which can be modeled to determine systematic deformations common for each image. A typical set of additional parameters can include radial lens distortion, expressible by a polynomial, tangential lens distortion, and film deformations (affinity, skewness).

The error equations can, for example, be augmented by the following terms:

$$v'_{x_i} = \dots a_1 x'_i \left(r'_i - r'_o \right) + a_2 x'_i \left(r'^3_i - r'^3_o \right) + a_3 x'_i \left(r'^5_i - r'^5_o \right) + a_4 x'_i \left(r'^7_i - r'^7_o \right)$$

$$+ a_5 x'_i \cos 2\alpha_i + a_6 x'_i \sin 2\alpha_i - a_7 y'_i \cos 2\alpha_i - a_8 y'_i \sin 2\alpha_i - a_9 x'_i \dots$$

$$v'_{y_i} = \dots a_1 y'_i \left(r'_i - r'_o \right) + a_2 y'_i \left(r'^3_i - r'^3_o \right) + a_3 y'_i \left(r'^5_i - r'^5_o \right) + a_4 y'_i \left(r'^7_i - r'^7_o \right)$$

$$+ a_5 y'_i \cos 2\alpha_i + a_6 y'_i \sin 2\alpha_i + a_7 x'_i \cos 2\alpha_i - a_8 x'_i \sin 2\alpha_i - a_9 y'_i + a_{10} x'_i \dots$$

α_i determines the axis of symmetry for asymmetric and tangential lens distortion and $r'_i = \sqrt{x'^2_i + y'^2_i}$ the radial distance; r'_o is a chosen fixed value, for example, 100 mm, to which radial distortion is referred.

A solution is first attempted without the use of additional parameters obtaining a particular σ_o for the solution. Thereafter, the parameters will be included. This should reduce σ_o. It is advantageous to statistically test the additional parameters to judge whether they contribute to the improvement.

For GPS flights, when observations for the exposure stations are added, further additional parameters modeling the effect of cycle slips between strips may be included. The bundle block adjustment is a rigid solution, which can be applied to any photogrammetric orientation and point determination problem. However, some simpler approaches are applicable for the orientation of single stereo models, such as relative and absolute orientation.

Relative Orientation

Relative orientation of a photograph with respect to the first photo is made if the exterior orientation of the first photo to ground coordinates (e.g., by space resection) has not yet been accomplished. Therefore, one makes the choice of local object coordinate system, the model coordinate system, which is based on the assumption that the first image is a vertical image with $x'_o, y'_o, z'_o, \omega', \varphi', \kappa'$ all zero. The calculation of the orientation parameters of the second photo, $x''_o, y''_o, z''_o, \omega'', \varphi'', \kappa''$, needs to be determined, so that stereoscopic viewing along epipolar lines is possible. The determination of x''_o is not yet of importance, since the model coordinate system has an arbitrary scale.

Figure 3.71 shows that the vectors $b', \lambda'_i p'_i$, and $\lambda'' p''_i$ form a triangle O', O'', and P. The addition of these vectors $b' = \lambda'_i p'_i - \lambda''_i p''_i$ requires the knowledge of the model coordinates formed by a space intersection. A solution is possible without these, using the vectors b', $\vec{p}'_i$, and $\vec{p}''_i$ only without the scale factors. The relations in the triangle may be expressed by the vector product of the three vectors, stating that the tetrahedron spanned between these vectors must be zero:

$$\left(b' \times p'_i\right) \cdot p''_i = 0$$

This is known as the coplanarity equation. As these vector operations may only be performed in one chosen coordinate system, for example, that of b', the components of p'_i and $\vec{p}''_i$ must be expressed in their projections to the axes of the base vector. Camera 1 has the rotational matrix, R', and camera 2, R'':

$$\vec{p}'_i = \begin{pmatrix} u'_i \\ v'_i \\ w'_i \end{pmatrix} = R' \begin{pmatrix} x'_i \\ y'_i \\ -f \end{pmatrix} = \begin{pmatrix} x'_i \\ y'_i \\ -f \end{pmatrix}$$

and

$$\vec{p}''_i = \begin{pmatrix} u''_i \\ v''_i \\ w''_i \end{pmatrix} = R'' \begin{pmatrix} x''_i \\ y''_i \\ -f \end{pmatrix}$$

The coplanarity equation may be written in determinant form with the components in the system of the base:

$$\begin{bmatrix} bx & by & bz \\ u'_i & v'_i & w'_i \\ u''_i & v''_i & w''_i \end{bmatrix} = 0 = \begin{bmatrix} bx & by & bz \\ x'_i & y'_i & -f \\ u''_i & v''_i & w''_i \end{bmatrix}$$

In order to obtain observation equations, linearization by differentiation and Taylor expansion becomes possible:

$$
\begin{vmatrix} bx & by & bz \\ 1 & 0 & 0 \\ u_i'' & v_i'' & w_i'' \end{vmatrix} v_{x_i}' + \begin{vmatrix} bx & by & bz \\ 0 & 1 & 0 \\ u_i'' & v_i'' & w_i'' \end{vmatrix} v_{y_i}' + \begin{vmatrix} bx & by & bz \\ x_i' & y_i' & -f \\ 1 & 0 & 0 \end{vmatrix} v_{x_i}''
$$

$$
+ \begin{vmatrix} bx & by & bz \\ x_i' & y_i' & -f \\ 0 & 1 & 0 \end{vmatrix} v_{y_i}'' + \begin{vmatrix} bx & by & bz \\ x_i' & y_i' & -f \\ \delta u_i'' & \delta v_i'' & \delta w_i'' \\ \hline \delta\omega'' & \delta\omega'' & \delta\omega'' \end{vmatrix} \delta\omega'' + \begin{vmatrix} bx & by & bz \\ x_i' & y_i' & -f \\ \delta u_i'' & \delta v_i'' & \delta w_i'' \\ \hline \delta\varphi'' & \delta\varphi'' & \delta\varphi'' \end{vmatrix} \delta\varphi''
$$

$$
+ \begin{vmatrix} bx & by & bz \\ x_i' & y_i' & -f \\ \dfrac{\delta u_i''}{\delta\kappa''} & \dfrac{\delta v_i''}{\delta\kappa''} & \dfrac{\delta w_i''}{\delta\kappa''} \end{vmatrix} \delta\kappa'' + \begin{vmatrix} 0 & dby & dbz \\ x_i' & y_i' & -f \\ u_i'' & v_i'' & w_i'' \end{vmatrix} - \begin{vmatrix} bx & by & bz \\ x_i' & y_i' & -f \\ u_i'' & v_i'' & w_i'' \end{vmatrix}_0
$$

The observation equations contain the observation errors in correlated form:

$$
\underset{j,i}{B}\, v_{i,1} = \underset{i,j\ j,i}{A\,x} - \underset{j,1}{l_{i,1}}
$$

For near vertical photography these correlations are, however, negligible, so that:

$$
Bv = b \cdot f\left(v_{yi}'' - v_{yi}'\right) = b \cdot f\left(py_i\right)
$$

$v_{yi}'' - v_{yi}'$ is the y-parallax, py_i, for which the error equations may be written as follows:

$$
v_{py}' = \frac{1}{bf} \left\{ \begin{aligned} & \left(-fu_i'' - x_i'w_i''\right)dby + \left(x_i'v_i'' - y_i'u_i''\right)dbz + \left[bx\left(y_i'v_i'' - fw_i'\right)\right. \\ & \left. dbz - by\left(x_i'v_i''\right) - bz\left(x_i'w_i''\right)\right]d\omega + \\ & \left[by\left(fw_i'' - x_i'u_i''\right) + bz\left(-fw_i''\right)\right]d\varphi'' + \left[bx\left(fu_i''\right) + by\left(fv_i''\right) + bz\right. \\ & \left. \left(x_i'u_i'' + y_i'v_i''\right)\right]d\kappa'' \end{aligned} \right.
$$

$$
\left. - \left\{ py_{\text{measured}} - \frac{1}{bf}\left[bx\left(y_i'w_i'' + fv_i''\right) - by\left(x_i'w_i'' + fu_i''\right)\right] + bz \right. \right.
$$

$$
\left. \left. \left(y_i'v_i'' + y_i'w_i''\right)\right] \right\}
$$

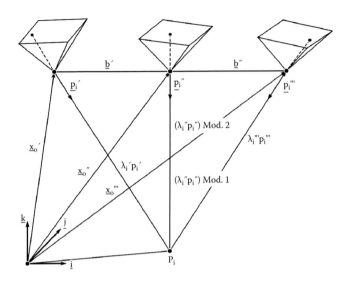

Figure 3.71   Scale transfer.

These error equations may be set up for the six von Gruber points, in which the y-parallaxes have been measured.

$$x_{5,1} = \left(\underset{5,6\ 6,5}{A^T A} \right)^{-1} \cdot \underset{5,6\ 6,1}{A^T \ell}$$

will yield in the orientation corrections db_y, db_z, $d\omega''$, $d\varphi''$, and $d\kappa''$ for the new photo. With these model coordinates of the orientation, points may be calculated by a space intersection with $bx = 1$ or with an arbitrary value.

It is possible to link the next stereo model to the first in this manner by a scale transfer, as shown in Figure 3.71.

Absolute Orientation

The next task is to orient the model with respect to the ground by a simple seven-parameter transformation. This process is called "absolute orientation" (see Figure 3.72).

The model coordinate system, x_i, y_i, z_i, has its origin at M and the ground coordinate system, X_i, Y_i, Z_i at G, G, M, and Pi form a triangle where the vectors ξ_o, $\lambda \cdot \vec{x}_i$, and ξ_i may be added to form:

$$\xi_i = \xi_o + \lambda \cdot T \cdot \vec{x}_i$$

$$\begin{pmatrix} X_i \\ Y_i \\ Z_i \end{pmatrix} = \begin{pmatrix} X_o \\ Y_o \\ Z_o \end{pmatrix} + \lambda \cdot T \begin{pmatrix} x_i \\ y_i \\ z_i \end{pmatrix}$$

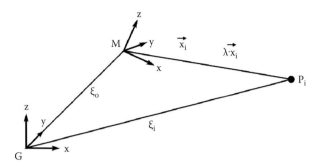

Figure 3.72   Absolute orientation.

T is a rotational matrix with the rotations Ω, Φ, and K for the absolute orientation. X_o, Y_o, and Z_o are the shift parameters to refer the coordinates to the origin of the ground system, and λ is the scale factor for the model.

Again error equations may be set up for the observed model coordinates. Their correlations may be neglected in absolute orientation:

$$\begin{pmatrix} v_x \\ v_y \\ v_{z_i} \end{pmatrix} = \frac{1}{\lambda} T^T \begin{pmatrix} X_i - X_o \\ Y_i - Y_o \\ Z_i - Z_o \end{pmatrix} - \begin{pmatrix} x_i \\ y_i \\ z_i \end{pmatrix}_{observed}$$

In linearized form, the equations become:

$$\begin{pmatrix} v_x \\ v_y \\ v_{z_i} \end{pmatrix} = \begin{pmatrix} \dfrac{\delta x_i}{\delta \lambda} d\lambda + \dfrac{\delta x_i}{\delta X_o} dX_o + \dfrac{\delta x_i}{\delta Y_o} dY_o + \dfrac{\delta x_i}{\delta Z_o} dZ_o + \dfrac{\delta x_i}{\delta \Omega} d\Omega + \dfrac{\delta x_i}{\delta \Phi} d\Phi + \dfrac{\delta x_i}{\delta K} dK \\[3mm] \dfrac{\delta y_i}{\delta \lambda} d\lambda + \dfrac{\delta y_i}{\delta X_o} dX_o + \dfrac{\delta y_i}{\delta Y_o} dY_o + \dfrac{\delta y_i}{\delta Z_o} dZ_o + \dfrac{\delta y_i}{\delta \Omega} d\Omega + \dfrac{\delta y_i}{\delta \Phi} d\Phi + \dfrac{\delta y_i}{\delta K} dK \\[3mm] \dfrac{\delta z_i}{\delta \lambda} d\lambda + \dfrac{\delta z_i}{\delta X_o} dX_o + \dfrac{\delta z_i}{\delta Y_o} dY_o + \dfrac{\delta z_i}{\delta Z_o} dZ_o + \dfrac{\delta z_i}{\delta \Omega} d\Omega + \dfrac{\delta z_i}{\delta \Phi} d\Phi + \dfrac{\delta z_i}{\delta K} dK \end{pmatrix}$$

$$- \begin{pmatrix} x_i - (x_i)_0 \\ (y_i - (y_i)_0) \\ (z_i - (z_i)_0) \end{pmatrix}$$

For near vertical photography, the relations can be simplified to:

$$v_x = \lambda \cdot x_i d\lambda + \lambda \cdot z_i d\phi + \lambda \cdot y_i d\kappa + dX_o - \lambda \cdot x_i + (X_0)_0$$

$$v_y = \lambda \cdot y_i d\lambda + \lambda \cdot z_i d\Omega - \lambda \cdot x_i d\kappa + dY_o - \lambda \cdot y_i + (Y_0)_0$$

$$v_z = \lambda \cdot z_i d\lambda - \lambda \cdot y_i d\Omega - \lambda \cdot y_i d\phi + dZ_o - \lambda \cdot z_i + (Z_0)_0$$

This equation system may be solved by a least squares adjustment to obtain the unknowns:

$$\lambda = \lambda_0 + d\lambda, \Omega_0 + d\Omega, \Phi = \Phi_0 + d\Phi, K = K_0 + dK, X_o = (X_o)_0 + dX_o,$$

$$Y_o = (Y_o) + dY_o$$

and $z_0 = (z_0)_0 + dZ_0$ to calculate the ground coordinates, X_i, Y_i, and Z_i of the model.

For the solution of absolute orientation of a stereo model, three control points forming a control point triangle are required (see Figure 3.44).

The application of this absolute orientation solution has also made it possible to perform an adjustment of aerial triangulation blocks for stereo models. For this, it is necessary to form spatial triangles between the adjacent von Gruber points of two models and the common perspective center of the exposure station. This solution was realized in F. Ackermann's Program for Aerial Triangulation Adjustment with Models (PATM) adjustment (see Figure 3.73).

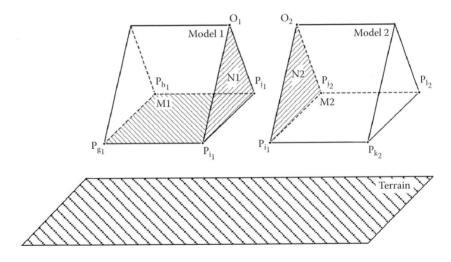

Figure 3.73 Model aerial triangulation.

DIGITAL PHOTOGRAMMETRIC OPERATIONS

The sequence of digital photogrammetric operations is described here with the help of the program package SIDIP (Simple Digital Photogrammetry), which originated at the University of Hannover. The core of the program is the bundle block adjustment (BLUH), programmed by K. Jacobsen of the university's Institute for Photogrammetry and GeoInformation. It was linked with semi-automatic photo measurement packages (DPLX by B. Pollak and DPCOR of the institute). The digital elevation model (DEM), orthophoto generation, and the GIS interface was provided by W. Linder of the Geographic Institute of the University of Düsseldorf (LISA-FOTO and LISA-Basic).

SIDIP

The program package Simple Digital Photogrammetry (SIDIP) is an example of a basic tool in digital photogrammetry. It does not possess the customized convenience for automated operations in a full production process, such as is contained in Z/I Imaging's Image Station 2001 or in LH-System's SOCET-SET, but it contains all of its basic elements. Alongside an explanation of SIDIP, references will be made here to these more powerful workstation solutions. SIDIP operates on a minimal configuration of a standard PC with 64Mb RAM and 5GB of storage in a standard Windows environment. It consists of five program parts, as shown in Figure 3.74.

DPLX

DPLX is a program for the semiautomatic measurement of points in images. Due to limitations in display storage of a full photogrammetric image, overview images are created in the form of image pyramids, which display the average of a specific number of pixel gray values (e.g., 4 × 4) as one pixel on the screen. A frame in the overview image can be selected to display a selected portion at full resolution. At the original image resolution, points may be selected for measurement. To more precisely select the pointing zoom images (e.g., 300%) may be created. At a click of a mouse, the measured image point will be marked with a cross and set at the appropriate point number (see Figure 3.75).

The movement of the measuring mark in the original image by the mouse is programmed for a fixed display of the image part.

More powerful commercial workstations can display the entire images directly, have a continuous zooming possibility, and permit roaming. During the roaming operation the measuring mark remains centered and the image is moved in relation to it. The measurement begins with the four or eight

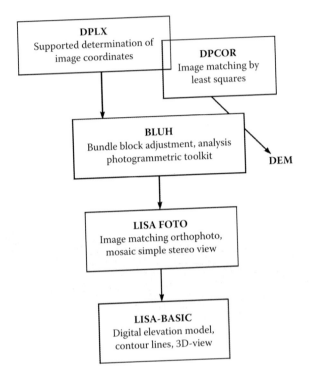

Figure 3.74   Program parts of SIDIP.

fiducial marks of the image. They permit the location of its principal point (see Figure 3.76).

The measurement is made in pixel coordinates $\vec{x}_i', \vec{y}_i'$. These can be transformed into image coordinates x_i', y_i' by the following relations

$$\begin{pmatrix} x_i' \\ y_i' \end{pmatrix} = \begin{pmatrix} \cos\kappa - \sin\kappa \\ \sin\kappa - \cos\kappa \end{pmatrix} \begin{pmatrix} \vec{x}_i' - \vec{x}_o' \\ \vec{y}_i' - \vec{y}_o' \end{pmatrix}$$

in which, for four fiducial marks 1 to 4, the principal point has the pixel coordinates

$$\vec{x}_o' = \frac{\vec{x}_1' + \vec{x}_2' + \vec{x}_3' - \vec{x}_4'}{4}$$

$$\vec{y}_o' = \frac{\vec{y}_1' + \vec{y}_2' + \vec{y}_3' - \vec{y}_4'}{4}$$

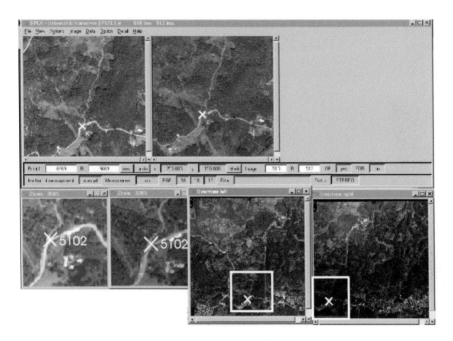

Figure 3.75   DPLX screen. (Image courtesy of the Institute for Photogrammetry and GeoInformation, University of Hannover, Germany.)

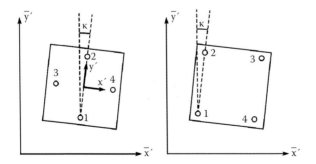

Figure 3.76   Measurement of fiducial marks.

or the transformation is carried out by an affine model

$$x_i' = a_1 + a_2 \vec{x}_i' + a_3 \vec{y}_i'$$

$$y_i' = a_4 + a_5 \vec{x}' + a_6 y_i'$$

and

$$tg\kappa = \frac{\vec{x}_2' + \vec{x}_1'}{\vec{y}_2' + \vec{y}_1'}$$

This process is called "interior orientation." For four or more fiducial marks, it may be subjected to a least squares adjustment with indication of the residuals.

The fiducial marks have round images in the photo. This permits them to be accurately measured automatically by an ellipse operator applicable to circular or ellipse-shaped targets. In the pixel grid covering the target, the gray level edges of the target are identified for each horizontal and vertical line of that grid. Their position is halved for all horizontal and vertical lines. These midpositions of the lines can be linked by two straight lines, the intersection of which results in the coordinates of the fiducial mark. DPLX includes not only the automatic measurement of fiducial marks but also the automatic measurement of signalized points or of symmetrical manholes (see Figure 3.77).

More advanced workstations have the added possibility of performing the entire interior orientation process automatically.

For measurements on fiducials and image points, an alphanumeric display permits the display of measured points in both photos with a listing of point number, pixel coordinates $\vec{x}_i'$, $\vec{y}_i'$, and image coordinates $\vec{x}_i'$, $\vec{y}_i'$. The listing shows the measured fiducial marks and the selected terrain points measured as control points and terrain points visible at least in two images or as transfer points visible in three images along a flight strip, or in six images across two strips.

Figure 3.77 Automatic measurement of fiducial marks. (Image courtesy of the Institute for Photogrammetry and GeoInformation, University of Hannover, Germany.)

DPCOR

The module DPCOR is integrated into DPLX for automatic image matching, which is based on a region growing method starting from seed points. DPCOR uses an image correlation for the determination of the approximate positions of corresponding points, which will be improved by least squares matching. Figure 3.78 shows the screen display of the equivalent ERDAS Imaging/Stereoanalyst software.

BLUH

The image coordinates computed by the interior orientation process are passed on into the bundle block adjustment program BLUH, consisting of more than 50 modules, the most important functions of which are the following:

- Preparation of photo coordinates permits the changing of point numbers or the numbers of groups of points; the sorting of the observations; and the correction of photo coordinates using calibration values

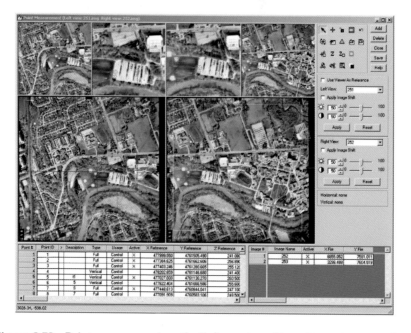

Figure 3.78 Point measurement on the Erdas system. (From Stereo imagery of Los Angeles, California; Data provider: HJW Inc.; Software tool: Imagine Ortho BASE® Erdas; Illustration provider: Erdas Inc., the Geographic Imaging unit within Leica Geosystems GIS & Mapping Division, Atlanta, Georgia.)

for lens distortion, earth curvature, refraction, and film deformations by transformation to the measured fiducial marks. The residuals of the fiducial mark measurements are indicated with a check of acceptability.

- Calculation of approximate photo orientations, the search for errors in the point measurements, and the sorting of observations to conform to a reduced bandwidth for the normal equations. With the aid of identical point numbers in the photos of a strip, object coordinates are calculated by two-dimensional transformation. Thereafter, adjacent strips treated in the same manner are linked in two dimensions. All computations are checked by data snooping to detect errors, which are indicated and eliminated.

- Setting of options for output files, the use of a particular choice of additional parameters, of error limits and weights for photo and ground coordinate measurements, the setting of robust estimator parameters for the adjustment, and the option to utilize additional measurements for in-flight GPS exposure stations and IMU exposure directions with appropriate weights.

- Execution of the least squares adjustment solution to determine the parameters of orientation, the measured points, and the additional parameters determined.

- Output in numerical form consists of lists of the observation errors, the achieved σ_o, and the object coordinates of all points measured as well as the image orientations.

- Output in graphical form shows the images and the discrepancy vectors with respect to the ground control points used. It is also possible to show the effect of additional parameters on the image coordinates (see Figure 3.79). The required geodetic coordinate conversions, in the case of high altitude and space images, are also included as an option.

- For scanner images, the standard geometric model using the perspective collinearity equations, needs to be replaced as an option. Instead of the collinearity equations, a separate scanner geometry is substituted. For a scanner, the image projection center x'_o moves forward from a beginning position, x'_{o_B}, to an end position, x'_{o_E}, in the image. The in-between projection center, x'_{o_i}, at distance b_i, can be interpolated along the base, b_{BE}:

$$x'_{o_i} = x'_{o_A} + \frac{b_i}{b_{BE}}(x'_{o_E} - x'_{o_B})$$

i represents the number of the scan line as a function of time.

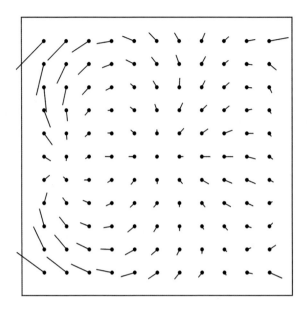

Figure 3.79   Image coordinate corrections determined by additional parameters.

An image point along the scan line, i, will therefore have the coordinates:

$$x'_i = (0, y', -f)$$

The imaged object point, x_i, along the line can thus be determined from:

$$X_i = \lambda \cdot R \cdot x'_i + x'_{o_i}$$

As a first approximation, the orientation matrix, R, and its parameters, φ, ω, κ, can be taken as constant for a satellite scanner image. A number of control points can ensure that the orientation changes during the platform motion are modeled as additional parameters.

For an aircraft scanner image, φ_i, ω_i, and κ_i, along the flight strip must be known as a function of time, t_i, through an IMU.

$$X'(t) = -f \frac{a_{11}(t)(X - X_0(t)) + a_{12}(t)(Y - Y_0(t)) + a_{13}(t)(Z - Z_0(t))}{a_{31}(t)(X - X_0(t)) + a_{32}(t)(Y - Y_0(t)) + a_{33}(t)(Z - Z_0(t))}$$

$$Y'(t) = -f \frac{a_{21}(t)(X - X_0(t)) + a_{22}(t)(Y - Y_0(t)) + a_{23}(t)(Z - Z_0(t))}{a_{31}(t)(X - X_0(t)) + a_{32}(t)(Y - Y_0(t)) + a_{33}(t)(Z - Z_0(t))}$$

This formulation may be utilized if the IMU records for $\omega(t)$, $(\varphi)t$, $\kappa(t)$ and the GNSS positioning values $x_o(t)$, $y_o(t)$, $z_o(t)$ have been recorded with sufficient accuracy. If this is not so, and all data need to be treated by a joint adjustment process, this formulation becomes too cumbersome.

For space imagery the rational function model has been introduced by the U.S. Department of Defense at a time of analytical plotters, when time-dependent panoramic photography was used. This model has now generally been adopted by the providers of space imagery who not only supply images in digital form but also the coefficients of rational polynomial functions (RPCs). These are calculated from the satellite's orbital position and orientation for the respective sensor model. Even a global DEM (e.g., from SRTM) can also be included for the determination of polynomial coefficients (Grodecki and Dial, 2003).

The RPC model relates object space coordinates x, y, z, which are functions of longitude, latitude, and elevation, back to image space coordinates via two cubic polynomials. Both object as well as image space coordinates are normalized for computational convenience.

Each scan line, imaged by a line sensor y' is a function of X, Y, Z:

$$Y' = \frac{P_1(X,Y,Z)}{P_2(X,Y,Z)}$$

with

$$
\begin{aligned}
P_1(X,Y,Z) =\ & a_1 + a_2 Y + a_3 X + a_4 Z \\
& + a_5 YX + a_6 YZ + a_7 XZ \\
& + a_8 Y^2 + a_9 X^2 + a_{10} Z^2 + a_{11} XYZ \\
& + a_{12} Y^3 + a_{13} YX^2 + a_{14} YZ^2 + a_{15} Y^2 X \\
& + a_{16} X^3 + a_{17} XZ^2 + a_{18} Y^2 Z + a_{19} X^2 Z \\
& + a_{20} Z^3
\end{aligned}
$$

$$
\begin{aligned}
P_2(X,Y,Z) =\ & 1 + b_2 Y + b_3 X + b_4 Z \\
& + b_5 YX + b_6 YZ + b_7 XZ \\
& + b_8 Y^2 + b_9 X^2 + b_{10} Z^2 + b_{11} XYZ \\
& + b_{12} Y^3 + b_{13} YX^2 + b_{14} YZ^2 + b_{15} Y^2 X \\
& + b_{16} X^3 + b_{17} XZ^2 + b_{18} Y^2 Z + b_{19} X^2 Z \\
& + b_{20} Z^3
\end{aligned}
$$

Each pixel along the scan line X' is also expressed as a polynomial function:

$$X' = \frac{P_3(X,Y,Z)}{P_4(X,Y,Z)}$$

with

$$P_3(X,Y,Z) = c_1 + c_2 Y + c_3 X + c_4 Z$$
$$+ c_5 YX + c_6 YZ + c_7 XZ$$
$$+ c_8 Y^2 + c_9 X^2 + c_{10} Z^2 + c_{11} XYZ$$
$$+ c_{12} Y^3 + c_{13} YX^2 + c_{14} YZ^2 + c_{15} Y^2 X$$
$$+ c_{16} X^3 + c_{17} XZ^2 + c_{18} Y^2 Z + c_{19} X^2 Z$$
$$+ c_{20} Z^3$$

$$P_4(X,Y,Z) = 1 + d_2 Y + d_3 X + d_4 Z$$
$$+ d_5 YX + d_6 YZ + d_7 XZ$$
$$+ d_8 Y^2 + d_9 X^2 + d_{10} Z^Z + d_{11} XYZ$$
$$+ d_{12} Y^3 + d_{13} YX^2 + d_{14} YZ^2 + d_{15} Y^2 X$$
$$+ d_{16} X^3 + d_{17} XZ^2 + d_{18} Y^2 Z + d_{19} X^2 Z$$
$$+ d_{20} Z^3$$

With a sufficient number of suitably located ground control points the 78 polynomial coefficients can be calculated and supplied to the user. To achieve the desirable local geocoding fit, the user may refine the RPC result by eventual offset corrections or by an affine transformation to local ground control.

LISA-FOTO

LISA-FOTO performs the tasks of a digital photogrammetric workstation. Figure 3.80 shows the user interface of the program.

The digitally stored photos used in DPLX and the photo orientations calculated in BLUH are imported into LISA-FOTO. This program has the following essential parts:

- Stereo measurement
- Stereo correlation
- Orthoimage generation

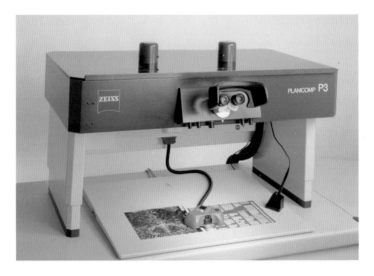

Figure 3.80 LISA-FOTO user interface. (Image courtesy of the Institute for Photogrammetry and GeoInformation, University of Hannover, Germany.)

Through the known image orientations from BLUH, the stereo model can be displayed in epipolar mode with no y-parallaxes present. Stereo viewing can be achieved in two simple modes: viewing of the images by a stereoscope or by anaglyphs.

The stereo measurement, like in DPLX, is initiated by selecting a model part on an overview image. Unlike DPLX, the program has roaming facilities in fast, intermediate, and slow motion, moving the images with respect to a fixed floating mark of changeable shape and color. The floating point can be kept at a constant height, or it can follow the elevation of a stored geocoded DEM. This is ideal for extracting vector information from the model since no changes of the floating mark are required during the plotting information.

Image Matching

Image matching or stereo correlation has a long history. In 1957, Gilbert Hobrough (Figure 3.81) patented an electronic image correlation. These initiatives sparked a number of further industrial developments by Raytheon-Wild (Stereomat B8), Bunker-Ramo-Wooldridge (Unamace), Bendix, and Gestalt to permit the automatic generation of digital elevation models by electronic image correlation. Through the present availability of high speed, large storage, digital computing these attempts have become obsolete.

Figure 3.81   Gilbert Hobrough, inventor of electronic image correlator, left, with Uki Helava (1959). (From the ISPRS Archives; previously published by ISPRS for the 100th Anniversary 2010 by G. Konecny.)

Digital image matching may be performed in different ways. The simplest method is the comparison of the cross-correlation coefficient between two images to be matched. A pattern matrix of a limited dimension (e.g., 17×17 pixels) of one image with gray values, d'_{ij}, can be compared with an equivalent size matrix of a second image with the gray values, d''_{ij}. For these two images, the variances σ', σ'', and the covariance r_{ij} may be calculated:

$$\sigma' = \sqrt{\frac{1}{n-1}\sum\left(d'_{ij} - d'\right)^2} \text{ with } d' = \frac{1}{n}\sum_{1}^{n} d'_{ij}$$

$$\sigma'' = \sqrt{\frac{1}{n-1}\sum\left(d'_{ij} - d''\right)^2} \text{ with } d'' = \frac{1}{n}\sum_{1}^{n} d''_{ij}$$

$$\text{cov}\, r_{ij} = \frac{1}{n}\sum d'_{ij} \cdot d''_{ij}$$

The cross-correlation coefficient, r_{ij}, becomes:

$$r_{ij} = \frac{\text{cov}\, r_{ij}}{\sigma' \cdot \sigma''}$$

This calculation is repeated within the search area larger than the kernel of the pattern matrix and the first search matrix by shifting the search matrix from left to right and from up to down. The result will be a matrix of cross-correlation

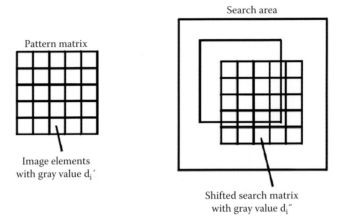

Search area

Pattern matrix

Image elements
with gray value d$_i$ ´

Shifted search matrix
with gray value d$_i$ ˝

Figure 3.82   Calculation of cross-correlation coefficients.

coefficients and their largest value will correspond to the x and y shift of the best match. The sequence of matches in x and then in y direction should result in a curve for the correlation coefficient with a maximum value. The search configuration is shown in Figure 3.82.

If the maximal correlation coefficient is greater than 0.7, a good match has been found. Depending on the contrast, the illumination differences, or the slope of the terrain, a significantly lower correlation coefficient may be obtained or the correlation might be lost.

One strategy to find the required approximate positions and to avoid loss of correlation is to work with image pyramids averaging 2×2, 4×4, 8×8, and 16×16 pixels into one density value. The correlation may start with the coarsest pyramid overview image. This will give approximations to start correlation at the next finer pyramid until the maximum correlation is reached at the highest pyramid level (see Figure 3.83).

Another strategy is to apply least squares matching. The aim of the method is to minimize the square sum of gray level differences between the pattern matrix and the geometrically transformed search matrix. In a first approximation, the transformation can be affine, or at a more refined level it can be projective. In image correlation, each density value, $d_i'(x,y)$, of the pattern matrix should correspond to the density value, $d_i''(x,y)$, of the search matrix so that:

$$d_i'(x,y) - v_i(x,y) = d_i''(x,y)$$

Assuming an affine deformation of the search matrix, the equation may be expanded

$$d_i'(x,y) - v_i(x,y) = r_o + r_1 \cdot d_i''(x',y')$$

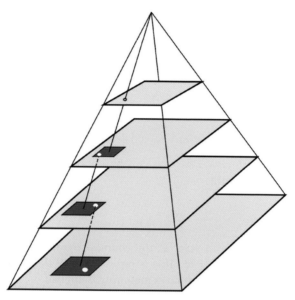

Figure 3.83   Image pyramids.

with

$$x' = a_o + a_1 x + a_2 y$$

$$y' = b_o + b_1 x + b_2 y$$

The linearization of the equation results in:

$$d_i'(x,y) - v_i(x,y) = d_i''(x,y) - d_o''(x,y)$$

$$+ \frac{\delta d''(x,y)}{\delta x} da_o$$

$$+ \frac{\delta d''(x,y)}{\delta x} da_1$$

$$+ \frac{\delta d''(x,y)}{\delta x} da_2$$

$$+ \frac{\delta d''(x,y)}{\delta y} db_o$$

$$+ \frac{\delta d''(x,y)}{\delta y} db_1$$

$$+ \frac{\delta d''(x,y)}{\delta y} db_2$$

The approximation value, $d_o''(x,y)$, is calculated with the assumptions

$$a_o = a_2 = b_o = b_2 = r_o = 0$$

and

$$a_1 = b_x = r_1 = 1$$

This leads to the observation matrix:

$$\ell_{n,1} + v_{n,1} = A_{n,u}X_{u,1}$$

$X_{u,1}^T = da_o, da_1, da_2, db_o, db_1, db_2, r_o, r_1$ can be obtained from a least squares solution $x = (A^TA)^{-1}A^Tl$ with:

$$\sigma_o = \sqrt{\frac{v^Tv}{n-1}}$$

In practice, correlation can start at the coarsest pyramid by calculation of the cross-correlation coefficient. This is repeated until the finest pyramid is reached. Following that, least squares matching is applied.

The correlation algorithms discussed thus far have to be executed for every pixel of the overlapping image matrix in which the pattern matrix (and the search matrix) is systematically moved around in horizontal and vertical direction. This leads to a time-consuming computation, which depending on pixel size, can take from minutes to hours on a small PC.

To economize on computation time, a selection of the measurement effort can be made. It is possible to correlate only in a regular point grid with the intent to interpolate elevation values later.

To narrow the extent of the search area, a regular object grid is chosen for which the location of the pattern matrix center and the search matrix center is calculated at different plausible elevations. For these locations the elevation corresponding to the highest cross-correlation coefficient is chosen as the correlation result. This procedure is utilized in LISA-FOTO. DPCOR can attempt to improve the obtained correlation result by applying least squares matching.

Another possibility is the selection of points, where the correlation is to be performed by an interest operator, which selects from all the image pixels those that are most suitable for an accurate correlation because of their symmetrical and large enough contrasts.

One of these operators is the Moravec operator. It calculates the mean square gradient, V_1, V_2, V_3, V_4, of a pixel window, $n \times n$, in all four main directions:

$$V_1 = \frac{1}{n(n-1)}\Sigma\Sigma[d(i,j) - d(i,j+1)]^2$$

$$V_2 = \frac{1}{n(n-1)} \Sigma\Sigma[d(i,j)-d(i+1,j)]^2$$

$$V_3 = \frac{1}{n(n-1)^2} \Sigma\Sigma[d(i,j)-d(i,j+1)]^2$$

$$V_4 = \frac{1}{n(n-1)^2} \Sigma\Sigma[d(i,j)-d(i,j+1)]^2$$

The objective is to find a min (V_1, V_2, V_3, V_4).

Another suitable operator is the Förstner operator. It is based on the model that the transformed gray level vicinity of a point, $d'(x, y)$, is a function of the original image, $d(x + x_o, y + y_o)$. This gives rise to an error equation:

$$d'(x,y) = d(x + x_o, y + y_o) + (x,y)$$

After linearization, setting the starting values, $x_o, y_o = 0$, this equation becomes

$$dd(x,y) - v(x,y) = \frac{\partial d}{\partial x} x_o + \frac{\partial d}{\partial y} y_o$$

with

$$dd(x,y) = d'(x,y) + d(x,y)$$

This leads to a normal equation matrix, N:

$$\begin{pmatrix} \Sigma\left(\frac{\partial g}{\partial x}\right)^2 & \Sigma\left(\frac{\partial g}{\partial x}\frac{\partial g}{\partial y}\right) \\ \Sigma\left(\frac{\partial g}{\partial y}\frac{\partial g}{\partial x}\right) & \Sigma\left(\frac{\partial g}{\partial y}\right)^2 \end{pmatrix} \begin{pmatrix} x_o \\ y_o \end{pmatrix} = \begin{pmatrix} \frac{\partial g}{\partial x} \cdot dd(x,y) \\ \frac{\partial g}{\partial y} \cdot dd(x,y) \end{pmatrix}$$

or

$$A^T A x = A^T(dd(x,y)) = Nx$$

This matrix, N, can be rotated to obtain an uncorrelated (diagonal) eigenvalue matrix, which permits the calculation of two characteristic values, w and q, for an error ellipse, so that

$$w = \frac{1}{\lambda_1 + \lambda_2}$$

and

$$q = 1 - \left(\frac{\lambda_1 - \lambda_2}{\lambda_1 + \lambda_2} \right)^2$$

w represents the size of the ellipse and q its roundness.

A point selected by the Förstner operator should have a minimum value, W_{min}, of

$$W_{min} = 0.5 \text{ to } 1.5(W_{\text{average for the entire image}})$$

and

$$q_{min} = 0.5 \text{ to } 7.5$$

The choice of interest operator points significantly reduces the image matching effort for the entire model. Other approaches for image matching have been suggested using extraction of line features in both images and in subsequent relational matching of the line features.

Image matching is usually done in image space. A resampling of the correlated pixel result into object space by resampling is then required. If the chosen correlation algorithm works in object space, it will in practice only provide a limited number of object points correlated, and the creation of an object pixel-based digital elevation matrix is also needed.

Semiglobal Matching

Image Matching software solutions were introduced by John Sharp of IBM in 1965, after an analogue electronic image correlator for the photogrammetric restitution was built in the 1950s by Gilbert Hobrough. Sharp introduced digital matching by area-based matching to derive a digital elevation model, which he used to produce an orthoimage. Figures 3.84 and 3.85 show Sharp and his early attempts in 1965 at IBM.

In area-based matching an image template in the first image is compared with an identical template within a search window over the second image. The template in the search window is moved around, and for each move the correlation coefficient is computed from all corresponding gray values of the pixels. The search window is always larger than the template. The gray level dependent sizes of template and search window both determine a successful match. In overlapping aerial photos the maximum value of the cross-correlation coefficient between the two overlapping images determines the height of the imaged terrain.

The search may be optimized if it is carried out at various resolution levels created by an image pyramid or if it is carried out along epipolar lines. The

Figure 3.84   John Sharp, digital image correlation at IBM, 1964, United States. (From the ISPRS Archives; previously published by ISPRS for the 100th Anniversary 2010 by G. Konecny.)

Figure 3.85   First automatic image correlation result by John Sharp. (From the ISPRS Archives; previously published by ISPRS for the 100th Anniversary 2010 by G. Konecny.)

epipolar lines on the photos are the trace of the plane formed between object point and both exposure stations.

Another search optimization possibility is provided by a search along a vertical line for a ground point x, y, while the elevation z is iterated. The ground points are in this case usually chosen for a regular grid in the object space.

In least squares matching, the change of geometry of the template is included as an unknown (e.g., as an affine deformation for one of the templates).

Feature-based matching for optimally filtered points or for edges detected has proven to be very useful in matching of automatically created transfer points for aerial triangulation.

Both area-based and feature-based matching approaches have faced the difficulty that they do not perform satisfactorily in areas with occlusions due to buildings in urban areas or due to vegetation in the forests.

The computer vision and pattern recognition community has, however, had a breakthrough in suggesting semiglobal matching as a possible solution (Hirschmüller, 2005). While dynamic programming techniques can enforce constraints within individual scan lines, global approaches in all directions, like graph cuts, can enforce matching constraints in two dimensions. Semiglobal matching is based in pixelwise matching with 2D smoothness constraints. These help to locate occlusions, which are given a lesser weight. The result is a better matching for areas with occlusions, such as urban areas.

Digital Elevation Models

The result of image matching is the creation of elevation data, z_i, at a specified object location, x_i, y_i. Depending on the correlation algorithm, these are located either in a regular grid (as in LISA-FOTO) or spaced in an irregular distribution (as by use of interest operators). In both cases an interpolation of these heights to a digital elevation matrix related to output pixels in the object system is required.

The interpolation can be done using the already determined height points in a stochastic field (see Figure 3.86).

P_i are the points with known elevations, z_i, in the vicinity of point, P_j, where z_j is to be interpolated. Interpolation can be done using a weight function, P_{ij}, which is proportional to the distance between P_i and P_j:

$$p_{ij} = \frac{1}{\sqrt{(x_j - x_i)^2 + (y_j - y_i)^2}}$$

$$z_j = \frac{\sum p_{ij} z_i}{\sum p_{ij}}$$

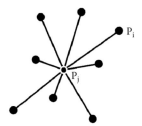

Figure 3.86 Interpolation in a stochastic field.

As weight function the square of the distance may also be used:

$$p_{ij} = \frac{1}{\left(x_j - x_i\right)^2 + \left(y_j - y_i\right)^2}$$

To consider a directional dependence, the location z_j can be expressed on an oblique plane with:

$$z_j = a_o + a_1 x_j + a_2 y_j$$

The coefficients, a_o, a_1, a_2 can be calculated as an adjustment problem from the nearest points, P_i, with the weights, p_{ij}.

The interpolation model can finally be expanded into least squares interpolation, which determines the distance dependent covariance function from all observed points and uses it as a weight to obtain z_j. Its advantage is to permit smoothing of the observation (Kraus, 1972). Another approach is the interpolation on the basis of finite elements (Ebner, 1979). Finally, it becomes possible to interpolate height on the basis of triangulated irregular networks (TINs) (Figure 3.87).

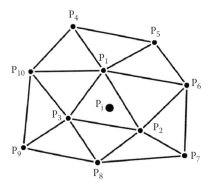

Figure 3.87 TIN interpolation.

Between all points at which elevations were measured, distances can be calculated. Three of these distances can be combined into a particular triangle. The triangle with the smallest area is finally chosen for the TIN, which consists of an assembly of smallest triangles generated.

For the triangle $P_1P_2P_3$ with its coordinates (x_1y_1, x_2y_2, x_3y_3) and heights (z_1, z_2, z_3) in which interpolation is to take place, z_i is expressed on the plane formed by the edges of the triangle:

$$z_i = a_o + a_1 x_i + a_2 y_i$$

The coefficients, a_o, a_1, a_2, can be calculated from the three equations for that triangle:

$$\begin{pmatrix} 1 & x_1 & y_2 \\ 0 & x_2 & y_2 \\ 0 & x_3 & y_3 \end{pmatrix} \begin{pmatrix} a_o \\ a_1 \\ a_2 \end{pmatrix} = \begin{pmatrix} z_1 \\ z_2 \\ z_3 \end{pmatrix}$$

If modeled in this way, adjacent triangles will not represent a smooth surface, but they will be separated by sharp edges.

To permit a smooth transition between triangles, the triangle surface may be expanded into a polynomial:

$$z_i = a_o + a_1 x_i + a_2 x_i^2 + a_3 x_i^3 + a_4 y_i + a_5 y_i^2 + a_6 y_i^3 + a_7 x_i y_i + a_8 x_i^2 y_i + a_9 x_i y_i^2$$

To solve for the coefficients of this polynomial, not only the heights, z_1, z_2, z_3, are available as observations, but also the slopes and curvatures in x and y of the adjacent triangles formed with P_4 to P_{10}.

The following 18 observation equations can be used for the triangle.

- For the elevations:

$$z_1 = a_o + a_1 x_1 + \cdots a_9 x_1 y_1^2$$

$$z_2 = a_o + a_1 x_2 + \cdots a_9 x_2 y_2^2$$

$$z_3 = a_o + a_1 x_3 + \cdots a_9 x_3 y_3^2$$

- For the slope:

$$\frac{\partial z_i}{\partial x_1} = a_1 + 2a_2 x_1 + 3a_3 x_1^2 + a_7 y_1 + 2a_8 x_1 y_1 + a_9 y_1^2$$

$$\frac{\partial z_i}{\partial y_1} = a_4 + 2a_5 y_1 + 3a_6 y_1^2 + a_7 x_1 + a_8 x^2 + 2a_9 x_1 y_1$$

similarly for

$$\frac{\partial z_2}{\partial x_2}, \frac{\partial z_2}{\partial y_2}, \frac{\partial z_3}{\partial x_3}, \frac{\partial z_3}{\partial y_3}$$

- For the curvature:

$$\frac{\partial^2 z_1}{\partial x_1^2} = 2a_2 + 6a_3 x_1 + 2a_8 y_1$$

$$\frac{\partial^2 z_1}{\partial y_1^2} = 2a_5 + 6a_6 y_1 + 2a_9 x_1$$

$$\frac{\partial^2 z_1^2}{\partial x_1 \partial_1} = a_1 a_4 + 2a_2 a_4 x_1 + 2a_1 a_5 y$$

similarly for

$$\frac{\delta^2 z_2^2}{\delta x_2^2}, \frac{\delta^2 z_2^2}{\delta y_2^2}, \frac{\delta^2 z_2^2}{\delta x_2 \delta y_2}, \frac{\delta^2 z_3^2}{\delta x_3^2}, \frac{\delta^2 z_3^2}{\delta y_3^2}, \frac{\delta^2 z_3^2}{\delta x_3 \delta y_3}$$

The other triangles will have similar observation equations. They can be solved in a simultaneous least squares adjustment for the 10 or more coefficients selected.

The advantage of the TIN modeling is that, along the lines of a particular triangle, natural discontinuities in the form of break lines can be considered in the interpolation process. In case of a break line, the equations for slope and curvature between the triangles that have a discontinuity are simply omitted.

At the end of the interpolation process each output pixel, x_i, y_i, in the object coordinates will have its corresponding z_i value. Figures 3.88, 3.89, and 3.90 show the DEM processing procedures.

Orthoimage Generation

With x_i, y_i, z_i known for each output pixel, the calculation of an orthoimage now becomes a simple matter. For the gray level transfer from one of the images forming a stereo model, the collinearity equations are used to calculate the corresponding $x_i' y_i'$ position. From there the gray value, d_i', is transferred to the output pixel as d_i, either by nearest neighbor assignment, by bilinear interpolation, or by cubic convolution (see Figure 2.86).

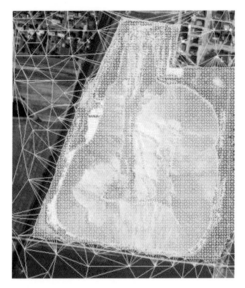

Figure 3.88 TIN structure of a DEM. (Image courtesy of the Institute for Photogrammetry and GeoInformation, University of Hannover, Germany.)

Figure 3.89 Interpolation of contours from the TIN. (Image courtesy of the Institute for Photogrammetry and GeoInformation, University of Hannover, Germany.)

Figure 3.90 Excessive slope calculation. (Image courtesy of the Institute for Photogrammetry and GeoInformation, University of Hannover, Germany.)

In the collinearity model, all image corrections determined from additional parameters $\Delta x'_i$ and $\Delta y'_i$ need to be included:

$$x'_i = -f \frac{r_{11}(x_i - x'_o) + r_{21}(y_i - y'_o) + r_{31}(z_i - z'_o)}{r_{13}(x_i - x'_o) + r_{23}(y_i - y'_o) + r_{33}(z_i - z'_o)} + \Delta x'_i$$

$$y'_i = -f \frac{r_{12}(x_i - x'_o) + r_{22}(y_i - y'_o) + r_{32}(z_i - z'_o)}{r_{13}(x_i - x'_o) + a_{23}(y_i - y'_o) + r_{33}(z_i - z'_o)} + \Delta y'_i$$

with

$$\Delta x'_i = a_1 x'_i (r'_i - r'_o) + \ldots - a_9 x'_i$$

$$\Delta y'_i = a_1 y'_i (r'_i - r'_o) + \ldots - a_{10} x'_i$$

(see Figure 3.79).

In the case of scanner images, the collinearity model needs to be replaced by the sensor model (see Figure 2.43 or Figure 2.47). Otherwise the procedure is the same.

Figure 3.91   Superimposition of vector information on the orthophoto. Stereo imagery of Los Angeles, California. (Data provider: HJW Inc.; Software tool: Imagine OrthoBASE® Erdas; Illustration provider: Erdas Inc., the Geographic Imaging unit within Leica Geosystems GIS & Mapping Division, Atlanta, Georgia.)

Adjacent orthophotos as part of different stereo models will have tone differences. To eliminate these, overlapping orthophoto areas may be created for which the orthophotos will be calculated from two photographs. The gray values from both of these object pixels may be taken as a mean. In this way orthomosaicking is accomplished.

The geocoded orthophoto coverage produced in this manner may be used directly for input and overlays in geographic information systems. The combination of digital orthophotos with vector information is possible by superimposition of raster and vector information. On the other hand, on-screen digitization with GIS software (MicroStation, Autocad, Geomedia, Arc-View or ArcInfo, etc.) becomes possible. Figure 3.91 shows an example of the superimposition of GIS information.

In the transfer of data, the format plays a role:

- Raw data, in which the gray levels of an image are stored uncompressed in a binary file, are not conducive to facilitate the use of the data. For color images there is the possibility of storing the pixels representing the three colors as either pixel interleaved (three pixels together), line interleaved, or band interleaved.

- The most common format is the TIFF data format, which contains a directory indicating image size, color dimension, and image resolution. GEOTIFF includes data on the geocoding parameters.
- The bitmap (BMP) format is suitable for 24-bit black-and-white and color images only for use in Microsoft Windows.
- The GIF format makes lossless data compression possible, however, at the expense of only 8 bits per pixel.
- The JPEG format has the best data compression capability. With JPEG, a matrix of 48 × 48 pixels is formed. The gray levels of the 48 × 48 pixel matrix, taking up much storage space, are not transferred as data but as a linear function, the coefficients of which are transferred. Thus, the volume of data is significantly reduced depending on the chosen compression level, but this cannot be done without loss.

A possible by-product of orthoimages are stereo orthophotos or stereo partners. If a DEM has been created, then these are easily computable. Orthophotos present a vertical view of the terrain. For the stereo partner, one can choose a particular constant viewing angle for creating x-parallaxes, Δx:

$$\Delta x_i = -\Delta z_i \cdot tg\alpha = -k \cdot \Delta z_i \text{ (see Figure 3.92)}$$

The stereo orthophoto can then be produced with the modified collinearity equations:

$$x_i' = -f \frac{r_{11}\left(x_i - k \cdot \Delta z_i - x_o'\right) + r_{21}\left(y_i - y_o'\right) + r_{31}\left(z_i - z_o'\right)}{r_{13}\left(x_i - k \cdot \Delta z_i - x_o'\right) + r_{23}\left(y_i - y_o'\right) + r_{33}\left(z_i - z_o'\right)}$$

$$y_i' = -f \frac{r_{12}\left(x_i - k \cdot \Delta z_i - x_o'\right) + r_{22}\left(y_i - y_o'\right) + r_{32}\left(z_i - z_o'\right)}{r_{13}\left(x_i - k \cdot \Delta z_i - y_o'\right) + r_{23}\left(y_i - y_o'\right) + r_{33}\left(z_i - z_o'\right)}$$

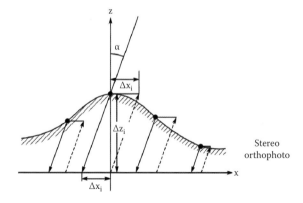

Figure 3.92   Stereo-orthophoto generation.

Orthophotos supply the correct geometry of a map and they have the advantage of offering visual interpretive capabilities. They do not, however, permit the display of object classes in the form of a GIS. Their use is therefore limited to acting as map substitutes for areas where timely mapping is too costly (urban areas) or too slow, or where map classification does not permit sufficient object details (swamps, coastal areas, forest) to be shown.

At a large-scale urban area orthophotos have particular problems: a digital elevation model attempts to record the elevation of ground points. An orthophoto produced with these elevations will, therefore, show radial displacements of buildings, trees, or bridges. Research has therefore been aimed at developing a methodology for the generation of so-called true orthophotos, in which the displacements for the building tops or bridge levels have been removed case by case.

Image correlation provides height information on the top of objects. An urban or a forested scene, to which image matching was applied, therefore generates a digital surface model rather than a digital terrain model.

SIDIP with the program RASCOR contains a filtering function by which elements not belonging to the surface may be ignored. This eliminates height measurements for building tops, and what remains is a DEM consisting of ground level heights.

ERDAS offers, with its packages Imagine and 3D GIS Stereo Analyst, the ability to vectorize, for example, a particular building in the orthophoto and to differentially remove its displacements in the orthophoto (see Figure 3.93).

The semantic modeling group of W. Förstner at the Institute for Photogrammetry of the University of Bonn has been successful in the 3D modeling of buildings. Edge matching is carried out for buildings, fitting the derived form to a particular building type, allowing the transformation of that building into an orthogonal view.

A. Gruen of the ETH Zürich and D. Fritsch of the University of Stuttgart and their groups use this strategy to derive 3D city models in which the façades contained in the images are rectified into the 3D scene.

LISA-Basic

LISA-Basic is a general digital elevation model program supported by break lines. Initially elevation zones are assigned gray levels. This raster DEM can be displayed as a raster black-and-white image or as a color-coded image for the zones chosen. It is also possible to interpolate contours, to display profiles, and to create oblique wire frame displays from these (see Figure 3.94).

Raster line data for contours may be converted into vector data for export in plot files in BMP, PCX, or DXF format. Finally, the orthophoto may be draped

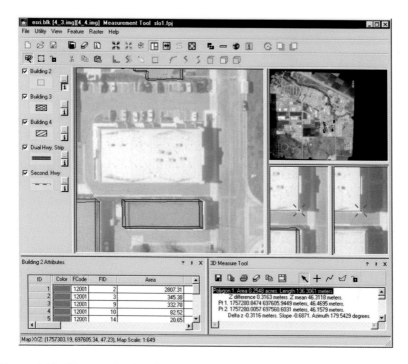

Figure 3.93 "True orthophoto" creation of ERDAS. Stereo imagery of Los Angeles, California. (Data provider: HJW Inc.; Software tool: Imagine OrthoBASE® Erdas; Illustration provider: Erdas Inc., the Geographic Imaging unit within Leica Geosystems GIS & Mapping Division, Atlanta, Georgia.)

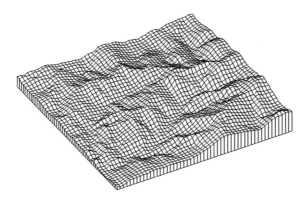

Figure 3.94 Wire frame display of a DEM. (Image courtesy of the Institute for Photogrammetry and GeoInformation, University of Hannover, Germany.)

Figure 3.95 Draped orthophoto projected onto a DEM. (Image courtesy of the Institute for Photogrammetry and GeoInformation, University of Hannover, Germany; Map courtesy of LGN, Hannover.)

over the oblique view of a DEM. The viewing position and the viewing direction can be arbitrarily chosen for the DEM. Then the ray between each DEM pixel and the viewing point is calculated and projected onto an image perpendicular to the viewing direction. The orthophoto gray level at the DEM pixel is transferred to the projected image. A resampling of this image is executed to obtain the oblique view (see Figure 3.95).

To omit hidden areas, the pixel-by-pixel calculation is carried out starting with the longest distance between DEM pixel and viewing point. For this ray, the gray level is transferred first. Then the calculation is repeated for the next shorter direction and so forth. In doing so the previous gray levels are overwritten, so that the hidden pixel gray values will not be visible in the oblique view.

LISA-Basic also has the capability to deduce secondary products of the DEM such as:

- Slope images, for which the slopes for a pixel are calculated between the adjacent eight pixels. The steepness of the slope, separated in different levels, can then be displayed in gray shades or in color code.
- Aspect images, for which the incident solar illumination at a certain hour (noon), latitude, and day of the year (declination) directed to a DEM pixel, determines an illumination vector. For an object point, a surface normal may be calculated from the adjacent elevation pixels. It is perpendicular to a plane expressed by x and y parameters. The angular difference between illumination vector and the vector of the surface normal may again be expressed in gray values or in color code for different magnitude levels of that spatial angle.
- It is furthermore possible to compare two different DEMs for the purpose of error analysis or for change detection, and to display the differences in an appropriate form with the described possibilities.

Racurs PHOTOMOD

SIDIP and LISA are university-produced programs which easily permit the photogrammetric restitution process as an educational experience (both programs are downloadable from the Web site www.ipi.uni-hannover.de for a small number of images).

Considerably more convenient and semiautomated operations are possible with commercially produced programs, such as PHOTOMOD, available from the Racurs Company in Moscow in the Russian Federation. The functions of PHOTOMOD are explained in the following illustrations as an example for other commercial program solutions, such as Leica LPS, Erdas, Inpho, KLT, and Socetset.

Figure 3.96 gives an overview of PHOTOMOD's processing capabilities: aerial photos, satellite images, and radar images may be processed into the contours of the digital terrain model (DTM); 2D and 3D vectors may be generated, and orthophotos and color balanced orthophoto maps may be created.

The operations start with the loading of the overlapping images (Figure 3.97). Originally a correlation-based algorithm was used to search for corresponding pixels in the overlapping images, every 10 to 100 m apart on the ground (Figure 3.98). Now new correlation based algorithms permit creation of a dense and nearly ground sampling distance (GSD)-based digital surface model (DSM)

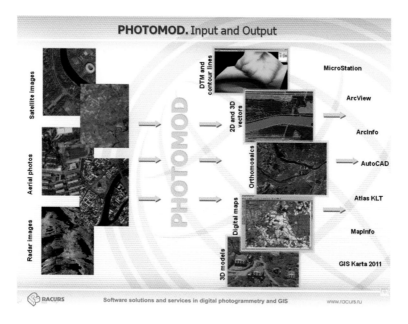

Figure 3.96 Racurs Photomod Processing capabilities. (From Racurs, Moscow, Russian Federation.)

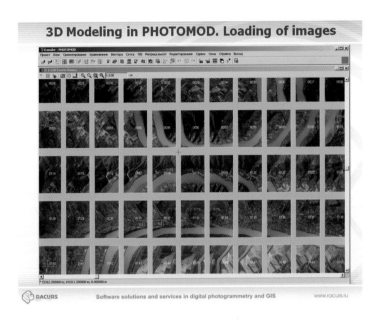

Figure 3.97 Loading of overlapping images. (From Racurs, Moscow, Russian Federation.)

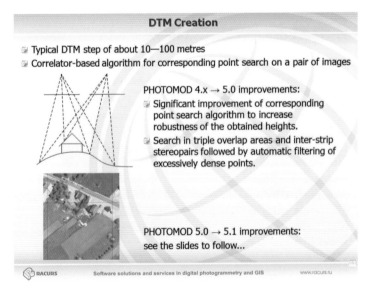

Figure 3.98 Coarse transfer point matching. (From Racurs, Moscow, Russian Federation.)

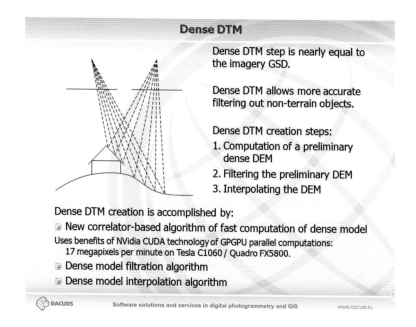

Dense DTM

Dense DTM step is nearly equal to the imagery GSD.

Dense DTM allows more accurate filtering out non-terrain objects.

Dense DTM creation steps:
1. Computation of a preliminary dense DEM
2. Filtering the preliminary DEM
3. Interpolating the DEM

Dense DTM creation is accomplished by:
- New correlator-based algorithm of fast computation of dense model
  Uses benefits of NVidia CUDA technology of GPGPU parallel computations:
  17 megapixels per minute on Tesla C1060 / Quadro FX5800.
- Dense model filtration algorithm
- Dense model interpolation algorithm

RACURS   Software solutions and services in digital photogrammetry and GIS   www.racurs.ru

Figure 3.99 Dense transfer point matching. (From Racurs, Moscow, Russian Federation.)

(see Figure 3.99). Figure 3.100 shows the automatically selected tie points for the aerial triangulation bundle adjustment (see Figure 3.101). Both bundle adjustment and the image matching operations are the basis for the DTM creation and the orthophoto production (see Figure 3.102).

Figure 3.103 shows the TIN and contours interpolated from the DTM. Figure 3.104 shows the contours after smoothing operations. Figure 3.105 illustrates how difficult areas for image matching, such as vegetation, settlements, or water surfaces may be excluded by filters to obtain a higher accuracy DTM. Figure 3.106 shows that by use of the filters a higher quality DTM may be created in an iterative process.

Figure 3.107 shows that a seamless orthomosaic with radiometric adjustment is accomplished by the use of seamlines between overlapping images. The result is an orthomosaic, shown in Figure 3.108.

The overlapping stereo-images may be used for a manual 3D vectorization of objects (Figure 3.109). It is also possible to fit house templates for an automatic roof vectorization of buildings (Figure 3.110). The result is a 3D city model (Figure 3.111). Terrestrial images may be used to fit the texture information of the house façades by rectification procedures to the 3D city model (see Figure 3.112).

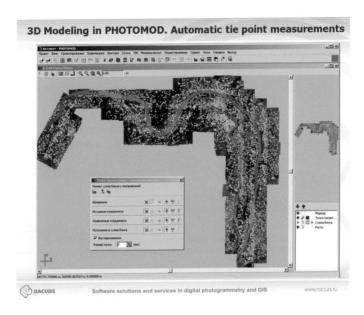

Figure 3.100 Dense matching carried out for aerial triangulation. (From Racurs, Moscow, Russian Federation.)

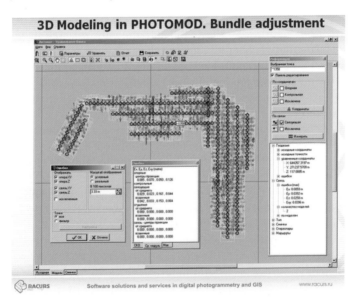

Figure 3.101 Automatic aerial triangulation. (From Racurs, Moscow, Russian Federation.)

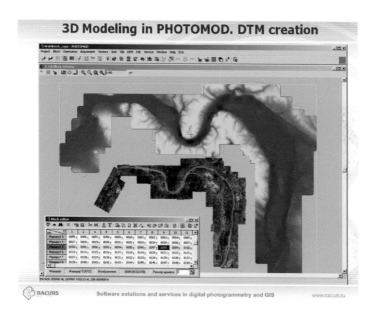

Figure 3.102   DTM creation. (From Racurs, Moscow, Russian Federation.)

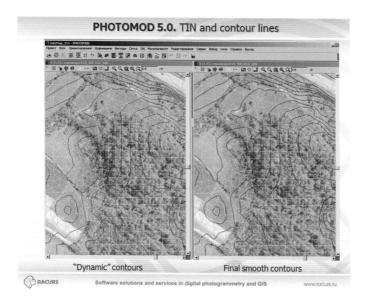

Figure 3.103   Interpolation of TINs and contours. (From Racurs, Moscow, Russian Federation.)

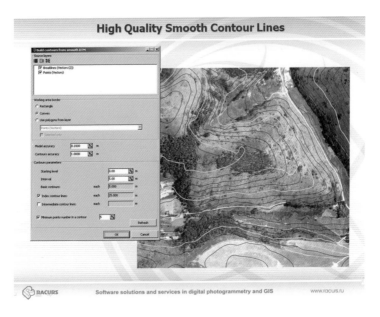

Figure 3.104   Smoothing of contours. (From Racurs, Moscow, Russian Federation.)

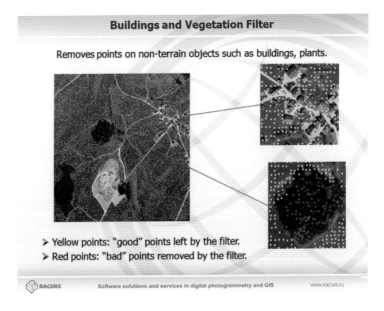

Figure 3.105   Filtering for areas with poor matching (vegetation, buildings). (From Racurs, Moscow, Russian Federation.)

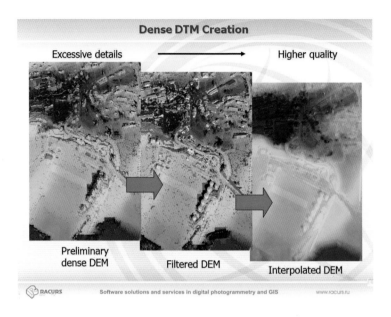

Figure 3.106 Filtering to match adjacent DTM parts. (From Racurs, Moscow, Russian Federation.)

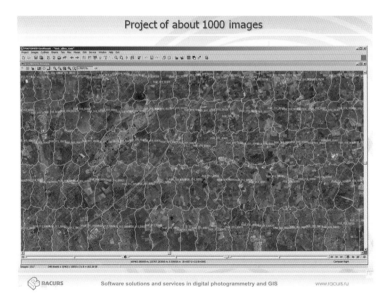

Figure 3.107 Seamline creation between adjacent orthoimages. (From Racurs, Moscow, Russian Federation.)

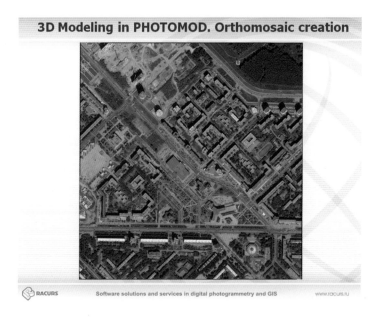

Figure 3.108   Orthomosaic. (From Racurs, Moscow, Russian Federation.)

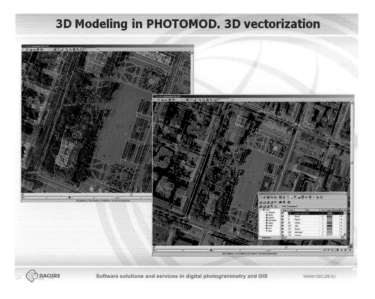

Figure 3.109   3D vectorization by operators. (From Racurs, Moscow, Russian Federation.)

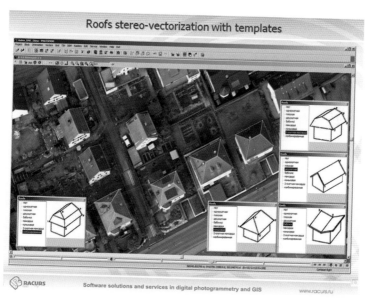

Figure 3.110 Fitting of roof templates for automatic roof vectorization. (From Racurs, Moscow, Russian Federation.)

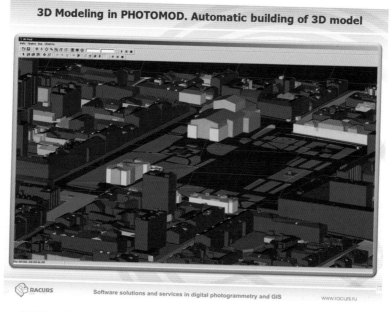

Figure 3.111 3D city model creation. (From Racurs, Moscow, Russian Federation.)

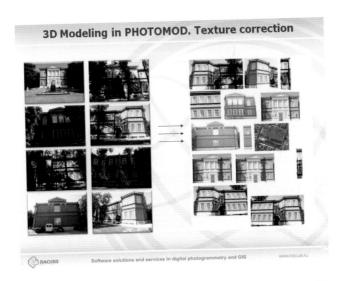

3D Modeling in PHOTOMOD. Texture correction

Figure 3.112 Fitting of terrestrial images to façades by rectification. (From Racurs, Moscow, Russian Federation.)

Of interest are also the programs for evaluating radar data. Figure 3.113 shows the RACURS application with images from TerraSAR-X (1 m GSD), Cosmo-Skymed (1 m GSD), and Radatsat-1 (25 m GSD), as compared to optical images of the same area (1 m GSD). Figure 3.114 illustrates the applications of radar imagery as an all-weather system for land cover classification, change detection, and radar interferometry for ground subsidence monitoring.

Figure 3.115 describes the task of coherent coregistration of subsequent SAR images. This is done on the basis of matching, generating phase-dependent parameters for these coherent images. Figure 3.116 shows the derivation of mean amplitude, amplitude stability, and mean coherence. Figure 3.117 depicts detailed images for the mean amplitude and for amplitude stability. In Figure 3.118, a detailed image of the mean coherence is shown; a pseudo-color representation for these three components permits a classification of the imaged terrain. Figure 3.119 shows the interpretation of the pseudocolors in the classification.

Figure 3.120 demonstrates that coherency permits one to provide information on moving objects for the image sequence. Figure 3.121 shows vehicle tracks in green.

Figure 3.122 describes the interferometric derivation of elevation differences from subsequent radar images (Uluru Rock, Australia).

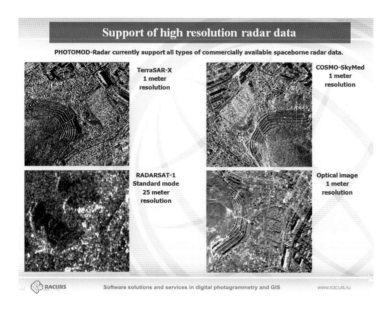

Figure 3.113 Racurs radar software for TerraSAR-X, Cosmo-Skymed, and Radarsat. (From Racurs, Moscow, Russian Federation.)

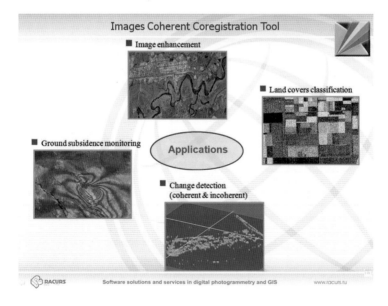

Figure 3.114 Applications of radar software. (From Racurs, Moscow, Russian Federation.)

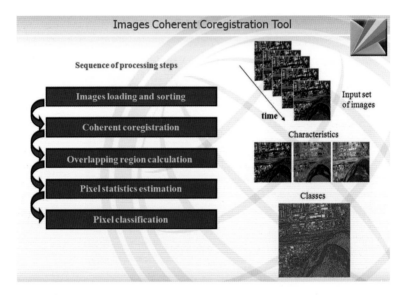

Figure 3.115   Coherent coregistration of subsequent radar images with phase dependent parameters. (From Racurs, Moscow, Russian Federation.)

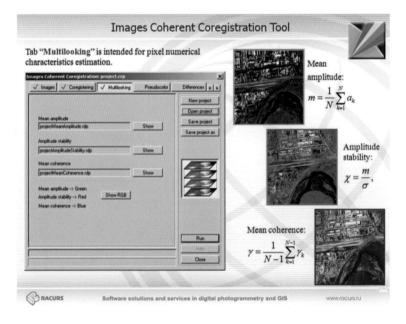

Figure 3.116   Mean amplitude, amplitude stability, and mean coherence. (From Racurs, Moscow, Russian Federation.)

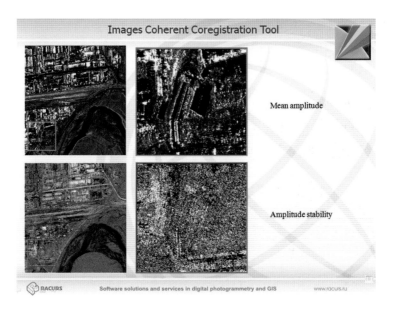

Figure 3.117 Image details of mean amplitude and amplitude stability. (From Racurs, Moscow, Russian Federation.)

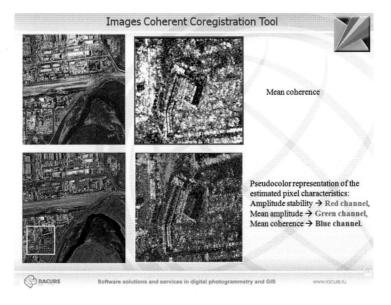

Figure 3.118 Image details of mean coherence and pseudocolor representation of the three components. (From Racurs, Moscow, Russian Federation.)

Name	Description	Color	Interpretation
Class 1	Low amplitude, high amplitude stability, low coherence.		Radar shadow.
Class 2	Low amplitude, high amplitude stability, high coherence.		Pavement.
Class 3	High amplitude, temperate amplitude stability, temperate coherence.		Buildings and metallic constructions. Front slopes of the hills and opposite slopes of ravines.
Class 4	High amplitude, high amplitude stability, high coherence.		Strong permanent scatterers.
Class 5	Low amplitude, low amplitude stability, temperate coherence.		Water surface.
Class 6	Low amplitude (less than for class 7), temperate amplitude stability, temperate coherence (higher than for class 7).		Vegetation.
Class 7	Low amplitude (higher than for class 6), temperate amplitude stability, temperate coherence (less than for class 6).		Water surface surroundings.

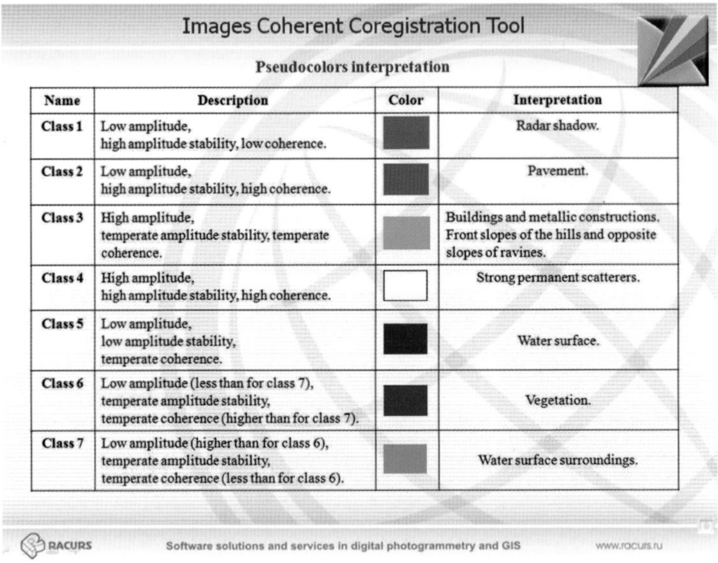

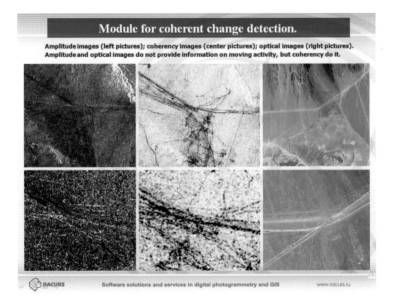

Figure 3.119 Interpretation of the pseudocolor representation to classify areas. (From Racurs, Moscow, Russian Federation.)

Figure 3.120 Coherency images detect moving objects. (From Racurs, Moscow, Russian Federation.)

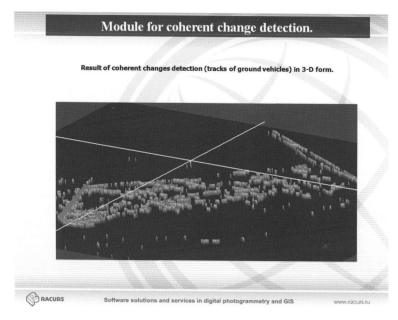

Figure 3.121 Vehicle track example. (From Racurs, Moscow, Russian Federation.)

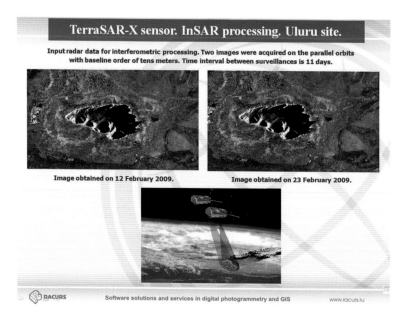

Figure 3.122 Interferometric derivation of elevation for Uluru Rock, Australia. (From Racurs, Moscow, Russian Federation.)

AIRBORNE LASER SCANNING

The term "laser" relates to its function: light amplification by stimulated emission of radiation. Lasers have found wide application in many disciplines. In remote sensing its first use was in 1972 with the Airborne Profile Recorder (APR), in which the heights between an airborne platform and the terrain were continuously recorded from laser pulse emissions. The APR had been used extensively in Canada since the 1950s.

In the 1980s, feasibility studies in the United States and Australia have led to the design of airborne laser scanners, which after 2000 introduced a worldwide market for laser scanning technology (Figure 3.123).

The basis of laser scanning is the measurement of the distance between aircraft and the ground, shown as vector d (Figure 3.124).

Of final interest is the 3D position shown as vector p:

$$p = f + d$$

Vector p must be simultaneously derived from the position vector of the aircraft f from GNSS measurements and from the measured laser distance d.

As GNSS measurements are typically only possible at a rate of 1 or 2 Hz, an inertial measurement unit IMU, which can receive rotational changes at the rate of 50 to 200 Hz, provides the possibility for an accuracy improvement to measure vector f by an interpolation process.

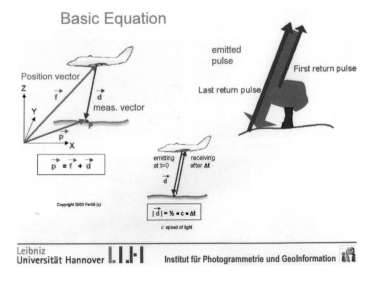

Figure 3.123 Principle of airborne laser scanning. (From the Institute for Photogrammetry and GeoInformation, Leibniz, University of Hannover, Germany.)

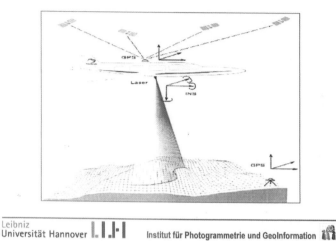

Figure 3.124   Laser scan during flight. (From the Institute for Photogrammetry and GeoInformation, Leibniz, University of Hannover, Germany.)

The advantage of the laser scanner does not lie in its capability to measure a single distance for vector *d*. As the laser measurement is possible at rates between 5000 and 25000 Hz, this offers the possibility to deflect the laser beam by a rotating mirror system. This makes it possible to cover the overflown terrain by a dense cloud of deflected laser pulse returns:

$$D = \tfrac{1}{2} \cdot c \cdot \Delta t$$

with Δt being the run time of the emitted pulse between emission and reflected return and c as the speed of light (299,792,458 m/sec).

To generate an optimally spaced point cloud on the overflown surface, different scanner manufacturers have used different scan patterns created by rotating, oscillating, or mutating mirrors for the deflection of the laser beam, for example, z-shaped scan or elliptical scan (Figure 3.125).

It is furthermore possible to record several returns for each emitted laser pulse. The first pulse received would therefore generally come from the DSM representing treetops or building tops, and a last pulse would be received from the ground, representing a DTM or DEM.

From the point cloud obtained by an airborne laser flight with an irregular pattern in general, a regularly spaced DSM, DTM, or DEM in a grid pattern tied to a coordinate reference system is interpolated for practical use.

Laser scanners use light in the range of $\lambda = 500$ to 1500 nm. The laser pulses are in the range between 1 and 10 nsec, and they are sent out in a narrow beam of smaller than 1 mrad.

Types of Scan - Pattern

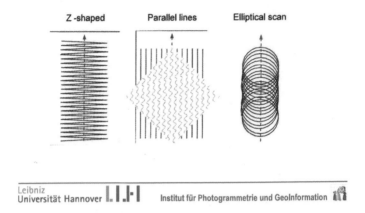

Figure 3.125 Different laser scan patterns used. (From the Institute for Photogrammetry and GeoInformation, Leibniz, University of Hannover, Germany.)

For vegetation classification a full-waveform lidar has been developed, for which multiple pulse returns are used.

Figure 3.126 shows the point cloud to the right and the laser intensity reflected image to the left. The gray values represent the intensity of the reflections. The intensity of reflections depends on the receiving power (P_R) which is a function of the distance (d), the beam width (D), the transmitted power (P_T), and the transmission coefficient (τ):

$$P_R = \tfrac{1}{4}\,(D/d)^2\,\tau\,P_T$$

This limits the altitude at which the airborne laser scanner can be operated to between 600 m and 4000 m altitude. With a chosen scan angle between 5 and 45 degrees, the achievable swath is between 300 and 2500 m in lateral direction of the flight. With a laser pulse rate of usually 150 Hz a point cloud consisting of up to 10 returns per square meter of the ground surface may be obtained.

The achievable height accuracy for a reflected point with airborne laser scanners is in the order of ±15 to 20 cm. Due to the complexity of determining

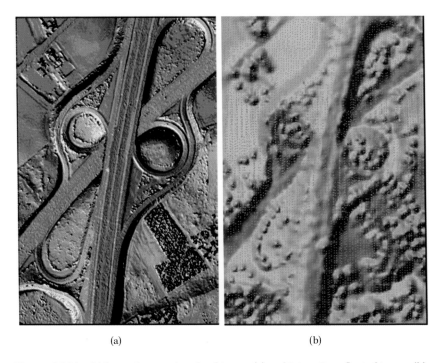

(a)   (b)

Figure 3.126   Airborne laser point cloud image (a) and intensity reflected image (b).

the position and the attitude of the scanner via GNSS and IMU the relative horizontal accuracy of a reflected point is slightly less, at about ±20 to 30 cm. The absolute position is often subjected to a bias of up to ±1 m or more, which must be corrected by ground control points. Contrary to contrast targets used for aerial photography, laser control points are best established as elevated targets.

Chapter 4

Geographic Information Systems

INTRODUCTION

A geographic information system (GIS), in a narrow definition, is a computer system for the input, manipulation, storage, and output of digital spatial data. In a more broad definition it is a digital system for the acquisition, management, analysis, and visualization of spatial data for the purposes of planning, administering, and monitoring the natural and socioeconomic environment. It represents a digital model of geography in its widest sense (see Figure 4.1).

In the narrow sense, a GIS consists of a system for data input in vector form, in raster form, and in alphanumeric form; a central processing unit (CPU) containing the programs for data processing, data storage, and data analysis and of facilities for visualization; and hard copy output of the data. In a broad sense, a GIS includes the data, which are managed by an administration or a unit conducting a project for the purposes of data inventory, data analysis, and data presentation for administrative support or for decision support.

The information system is based on data that are available in various forms:

- Spatial objects are represented by identifiers. They can relate to points, lines, or areas administered in vector form. The identification and organization of these objects in coordinate and vector form is subdivided into feature or object classes. This includes their spatial or topological relations in two or three dimensions.
- Data in raster form are also included. A pixel may be assigned an object code, or it may simply consist of gray levels of an image or a digital elevation model.

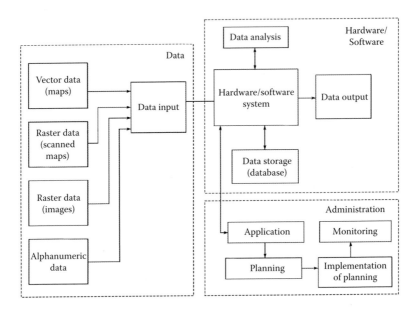

Figure 4.1   Concept of a geographic information system.

- The vector or the raster data are also linked to nongraphic information specifying place names and object numbers, which in databases may further be linked to a great variety of coded or alphanumerical attributes (e.g., owners of a parcel, inhabitants of a house, characteristics of a utility feature, statistical data for a defined area).

General and specialized GIS systems have been designed for a variety of purposes:

- Environmental management and conservation
- Defense and intelligence purposes
- Governmental administration
- Resource management in agriculture and forestry
- Geophysical exploration
- Cadastral management
- Telecommunications
- Utility management
- Business applications
- Construction projects

Many of these applications require common base data. It is the purpose of an administrative authority to create a spatial data infrastructure by which the

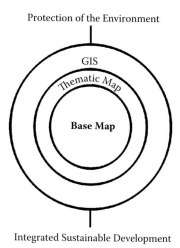

Protection of the Environment

Integrated Sustainable Development

Figure 4.2   The relationship between base data and thematic data.

base data may easily be exchanged. The main preoccupation in this context is the creation of a topographic database onto which thematic data of specific interest may be added (Figure 4.2).

The justification lies in the fact that cost–benefit studies of 39 GIS projects in Scandinavia in the 1990s, carried out by Nordisk Quantif, have shown that automation of a single production task in an administrative unit results in a cost–benefit ratio of 1:1, whereas the integration of data with other organizations can raise this ratio to 1:4 (Figure 4.3). This is all the more significant since, on average, the cost of data acquisition and the effort for its updating by far exceeds the costs for hardware, software, and data processing.

Throughout the development of GIS systems, the hardware costs steadily declined. After a rise in the 1980s, the software cost has likewise taken a downward turn (Figure 4.4). With increased hardware and software power, GIS and data management likewise get more efficient and cheaper. What remains high in cost is the provision of data, particularly if these are to be kept up to date to reflect a model of the actual geographic and socioeconomic environment.

The GIS pyramid of Figure 4.5 reflects the need for three types of users involved with GIS. Most users are involved solely in the viewing aspects of GIS data. A smaller group is involved in the analysis of data. This is done by modeling to arrive at information deduced by combining different GIS data sets. A small group of GIS users is involved in base, thematic, and attribute data provision and its updating.

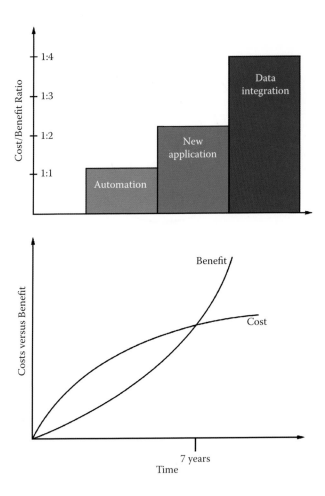

Figure 4.3   Cost–benefit ratio of GIS projects.

Figure 4.6 shows the evolution of a GIS system over a number of years. Initially, the focus is on providing an inventory of data. Later, the analysis is highlighted. Finally, the emphasis is on management.

HARDWARE COMPONENTS

The hardware of a GIS is composed of (Figure 4.7):

- Input devices
- Processing and storage devices
- Output devices

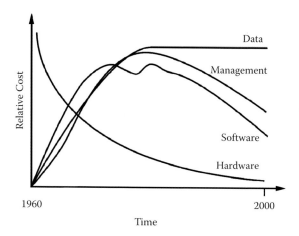

Figure 4.4   GIS cost aspects.

Input Devices

Digital data input depends on the type of data to be utilized.

Imagery input is possible from analogue images through the use of image scanners. Digital airborne and spaceborne systems already use charge-coupled device (CCD) sensors to supply the data in digital form. Light falling onto a semiconductor is transformed into an electric charge and into electric current. The light energy is proportional to the electric current, and thus brightness measurement becomes possible.

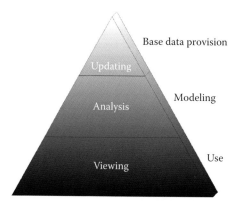

Figure 4.5   GIS pyramid.

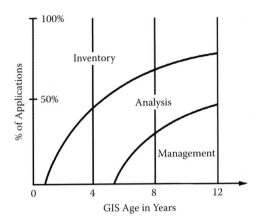

Figure 4.6   Evolution of a GIS.

Area CCDs are capable of providing a full frame transfer of a shutter released image at full resolution. However, they suffer from long read-out times. Future complementary metal–oxide semiconductor (CMOS) technology may overcome the current size limitations of area CCDs. High-resolution systems therefore prefer the use of long linear arrays operated as push-broom scanners, which integrate charges and read them out line by line, without the use of a shutter.

For analogue images a resolution of 50 lp/mm can be reached and for digital images even more.

Maps can be manually digitized by two-dimensional digitizers in vector form. This is possible in a single point mode or by dynamic measurement based on distance or time. The resolution of digitizing is about 0.2 mm. This is achieved by a fine wire grid inside the digitizing table. Digitizers are available for an A2 format or larger.

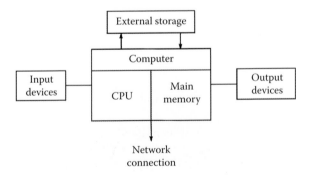

Figure 4.7   Hardware components.

Maps may, however, also be raster scanned using scanners. These are available as drum scanners or as flatbed scanners with a pixel size of 7 μm and an accuracy of 2 to 4 μm. They range from inexpensive desktop scanners limited in geometric and radiometric resolution to expensive but accurate cartographic scanners varying in geometric and radiometric resolution. They can be used in transparent or opaque mode, and are suitable for black-and-white or color scanning. Scanned raster data may subsequently be converted into vector information with GIS software.

3D-vector data can be obtained directly by terrestrial survey equipment, such as

- Theodolites
- Electronic tacheometers
- Leveling instruments
- GPS receivers
- Mobile mapping systems

3D information from aerial photographs may be compiled by analogue or analytical plotters or by digital photogrammetric workstations.

The advantage of manual 2D or 3D *vector digitization* is that nongraphic attribute data may easily be attached via keyboard or menu. When manually digitizing vector information from images, the following limitations of the human eye should be considered. The monoscopic acuity of the human eye at a viewing distance of 25 cm is about 7 lp/mm. For an imaging system with a resolution of 60 lp/mm, this permits a viewing magnification of about 8 for a 7 μm pixel size of image. Current computer screens with 1664 × 1248 pixels only have a resolution of 300 μm per screen pixel.

Considering that the stereoscopic acuity of the eyes is even better, this means that screen digitization is inferior to 3D digitization (line extraction) on analytical plotters. Digital photogrammetric workstations, however, offer advantages for automated operations.

Processing and Storage Devices

Processing and storage devices consist of the central processing unit (CPU) and the main memory, the external storage devices, and the user interface (see Figure 4.8).

The *CPU* executes the program commands. Its arithmetic unit performs algebraic and logical operations for the data. Its control unit regulates the data transfer between the arithmetic unit and the main memory.

The *main memory* (random access memory, or RAM) contains the machine programs and accepts data in short access time with caching, if required. The

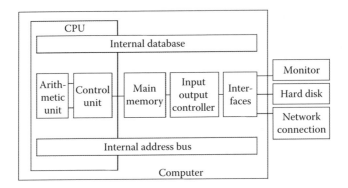

Figure 4.8   CPU and main computer memory.

I/O controller communicates with the periphery for hardware ports and for software drivers. The *bus system* establishes the connections. To speed up the output process, additional graphic cards and memory are usually added as interfaces.

Criteria for the CPUs performance are:

- Processor speed, 2 MHz in 1950 and 3 GHz after 2000
- Internal data format (32 or 64 bit)
- External data format between the CPU and the main memory (64 bit)
- Physical memory, in the range of megabytes in the 1980s, gigabytes in the 1990s, and terabytes in 2010
- Computing performance, 0.002 MIPS in 1950 (Univac), 10 MIPS in 2003 with the Pentium 4, and reached 150000 MIPS in 2010

External storage devices are linked to the computer. The following options are available:

- Magnetic hard disks of >30GB with <10 msec access
- For archival magnetic tapes, CD-ROMs for 650MB up to 3.2 GB are available with access speeds of <150 msec
- Jukeboxes are larger archival devices for a storage capacity of several terabytes (TB)

For the current GIS systems, the following computer configurations can be considered:

- PCs with one or several Intel Pentium Processors for use with Windows operating systems
- Workstations by
  - Hewlett-Packard, RISC
  - Sun, SPARC

- IBM, RISC
- DEC, Alpha
- SGI, MIPS

PCs or workstations are usually networked together in a client–server arrangement. The *server* holds the data in files or databases and contains application programs. The *client* is a user terminal (PC). The network is administered in a local area network (LAN) or in a wide area network (WAN).

The *user interface* consists of a high-resolution screen, which is adapted for color viewing and optionally for stereo. It also consists of a mouse and a keyboard.

Output Devices

Output devices include the ports to printers. Specific to GIS are the following graphic output facilities.

Vector devices are flatbed plotters and drum plotters. Flatbed plotters have an accuracy of ±0.05 mm at a speed of <30 m/min operated with a pen or a light beam. Drum plotters are less accurate but faster (300–900 mm/min). They are used for verification plots.

Raster devices permit the output of halftones in a pixel or a screened manner. They are able to print RGB or CYMK colors in different saturations. They can combine vector and raster data in raster form. To print halftones, the dithering technique is used, in which printer pixels are combined in halftone cells. For example, a 600 dpi output has a 150 dpi halftone cell for a 4-bit radiometric resolution.

Screened color reproduction uses different screen angles:

- 0° for yellow
- 15° for magenta
- 45° for black
- 75° for cyan

Laser raster drum plotters are available for film printing up to the A0 format with a resolution of 0.01 mm. Other output possibilities are by dye sublimation, thermal wax transfer, and inkjet technology or laser printing.

A typical configuration of a GIS hardware installation is shown in Figure 4.9.

SOFTWARE COMPONENTS

Operating Systems

The operation of a computer is based on its operating system. It ensures that all parts of the computer function in liaison. Most common are Microsoft's operating systems for PCs.

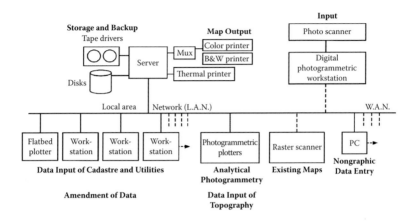

Figure 4.9   Typical GIS hardware installation.

In MS-DOS (Microsoft Disk Operating System) of 1978 the operation was regulated by text lines. This permits the administering of files by name. More modern *Windows* operating systems such as Windows 3.1, Windows 95, Windows 98, Windows NT, Windows 2000, Windows ME, Windows XP, Windows Vista, and Windows 7 and 8 utilize graphic symbols (icons). Windows acts as a graphical user interface (GUI). Windows is now a network compatible system.

An operating system for workstations was UNIX, which had been adapted for the computers of specific manufacturers: HP-UX by Hewlett-Packard, AIX by IBM, and Linux, which is available as an open source system. Unix development dates to the 1960s. It was originally designed for the operation of mainframe computers with multitasking. It contains a great number of data security features regulating access.

Programming Languages

The programming of computers is made possible by programming languages, which translate user formulations into machine-compatible code. For this translation a compiler for the respective programming language is required. Most GIS programs, based upon a chosen operating system, have been programmed in the programming language Fortran (formula translation). More modern languages are C, C++, and Visual Basic.

Networking Software

The communication of computers within a local area network (LAN) and a wide area network (WAN) is ensured by International Standards Organization

(ISO) standards. The most common standard is TCP/IP (Transmission Control Protocol/Internet Protocol). TCP/IP separates data transmission into smaller packages transmitted from an identified sender to a receiving computer. The transmission of the packages is checked during the process.

Graphic Standards

Graphic standards have been introduced so that the complex graphic instructions of the computer can be translated into monitor-compatible instructions. An internationally agreed graphic standard is the Graphical Kernel System (GKS). It defines 2D graphic primitives (position, height, line type, font, color, and fill). Other standards are X window (X11) and special standards for 3D graphic cards.

GIS Application Software

Based upon an operating system, augmented by additional programming tools and standards, various vendors (Esri, Intergraph, Siemens, and many others) have developed GIS software packages. They have a great number of elements in common:

- Translation (translation, rotation, and scale change in the two dimensions of the screen)
- Polygon creation (linking a line network to the origin of the line string)
- Adjustment of polygons (observing conditions of right angles and parallel lines)
- Line smoothing (connecting line strings by curves)
- Vector to raster conversion (for display of vector information)
- Raster to vector conversion (line derivation of vectors from pixels representing a line)
- Edge cut-off (to fit a seamless data set to the screen)
- Edge matching (to fit lines of adjacent tiles together)
- Geometry edit (to change point and line information)
- Intersections (between point locations, lines, and polygons) in one layer
- Intersections between layers
- Buffer zone generation for points, lines, and areas
- Counting of points in areas
- Measurement of point coordinates, distances, and areas
- Interpolation
- Modeling functions
- Network analysis
- Symbol and text generation

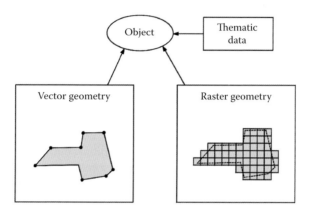

Figure 4.10   Vector and raster geometry.

- Generalization
- Map annotation

These tasks will be discussed for vector and for raster systems (Figure 4.10).

VECTOR SYSTEMS

Object Representation

The modeling of vector geometry depends on local or georeferenced coordinate systems. The advantage of vector systems lies in the possibility of recording and displaying coordinates with full measurement accuracy of ground surveys or of photogrammetric point and line measurements. In general, vector systems also contain less data volume than raster images of the same area. Furthermore, it is easy to attach alphanumeric attributes to the defined elements of a vector system (Figure 4.11), such as

- Points
- Lines
- Areas
- Objects

A point is defined by its coordinates, x, y, and by its node number. A line is defined by the coordinates of its end points, x_1, y_1, and $x_2 y_2$ and its line (arc) number. A line string is defined by the coordinates of all points forming the line string: $x_1 y_1, x_2 y_2, \ldots x_n y_n$. An area is defined by the coordinates of the line string ending at the initial points: $x_1 y_1, x_2 y_2, \ldots x_{n-1} y_{n-1}, x_1 y_1$. To points, lines, and areas, attributes with alphanumeric thematic data may be attached (Figure 4.12).

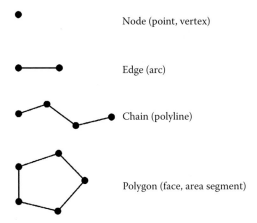

Figure 4.11   Point, line, and area objects.

In Figure 4.13a the different ways in which an area may be represented are shown. Figure 4.13b shows the digitization in the computer-aided design (CAD) form of "spaghetti graphics." CAD systems, such as Autocad or Microstation, in their simplest form do not automatically snap adjacent lines to common end and intersection points. Their result is a visual area representation, which cannot be analyzed for adjacency. Figure 4.13c shows the representation of areas by formation of closed individual polygons and by selection of additional line strings. Intergraph's MGE software provided this partial topology

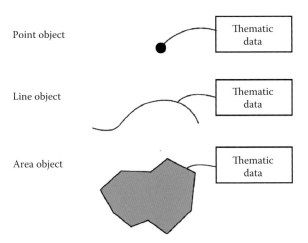

Figure 4.12   Attribute links.

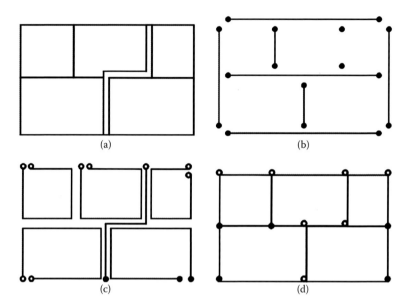

Figure 4.13 Graphic representation of an (a) area, (b) simple CAD lines, (c) closed individual polygons, and (d) polygons with full topology.

model. Figure 4.13d shows the generalized attempt to digitize areas through the formation of polygons and by intersecting them with lines. In this way, a relational geometry model may be built if the rules of topology are observed, as introduced in Esri's ArcGIS. The relations are expressed in three relational tables. These relate to Figure 4.14.

LINE–AREA TABLE		
Line	**Area Left**	**Area Right**
a	ι	φ
g	κ	φ
e	ι	φ
f	ι	φ
b	ι	κ
c	ι	κ
d	ι	κ
g	φ	κ

COORDINATE TABLE FOR ALL POINTS	
Point	**Coordinates**
1	$x_1 y_1$
2	$x_2 y_2$
3	$x_3 y_3$
4	$x_4 y_4$
5	$x_5 y_5$
6	$x_6 y_6$

LINE–POINT TABLE		
Line	**From Point**	**To Point**
a	1	2
b	2	3
c	3	4
d	4	5
e	5	6
f	6	1
g	2	5

Please note that the outside area is also identified, and that the direction of the line must be indicated to ensure full topology. The relational model may be contained in a large relational database (Oracle Spatial, Siemens) or it may be administered separately (ArcGIS and ArcSDI by Esri).

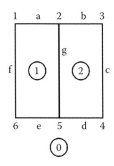

Figure 4.14   Topological model.

The knowledge of the topological relations permits neighborhood queries. Nongraphic attributes may be directly attached to the identifiers of points, lines, and areas or to objects composed of areas or line strings.

Early GIS developments favored the CAD or partial topology models, because they needed less computing power and could operate on larger databases. Attributes were then attached to "pointers," which consisted of points identifying the graphic placement of an area (parcel) number constituting the link to the nongraphic database content. In partial topology systems, the area number could directly establish the link to the attributes.

Since it was easier to digitize in CAD systems, topology was created by software snapping to common points and to line intersections in a postdigitizing batch process assuming a tolerance distance. In doing so, point numbers, line numbers, and area numbers were automatically created, and the relational tables were set up accordingly.

GIS systems adopted the *layer concept*, in which graphic information of a certain theme was stored in separate graphic files. These were easier to manipulate. Different layers could be superimposed in separate colors, much like the themes in a topographic map printed in different colors. The intersection of two different layers presented additional topology generation needs through a subsequent batch process. The generation of a GIS in an object-oriented manner is helpful in this. The German topographic system ATKIS groups the elements of topographic objects into hierarchical categories:

Code 1000, Control points
Code 2000, Buildings
Code 3000, Transportation
Code 4000, Vegetation
Code 5000, Hydrography
Code 6000, Topographic relief
Code 7000, Boundaries

Each category is divided into object groups, for example:

Code 3100, Roads
Code 3200, Railways
Code 3300, Air traffic
Code 3400, Ship traffic
Code 3500, Buildings for traffic purposes

Then each object group is divided into types, for example:

Code 3101, Road
Code 3102, Path
Code 3103, Square

Each object type, group, or category may contain its specific nongraphic attributes (Figure 4.15).

If the geometry of a defined object extends over several layers, we speak of an "object-oriented GIS." Attributes may be inherited for these objects, and specific processing methods may be applied to them. In the LH-Systems/Laser Scan Lamps 2 software, parts of the database in different layers may be commissioned by object orientation to different users.

Vector Geometry

The calculation of vector positions and intersections follows the principles of analytic geometry. Depending on the two- or three-dimensional representation chosen for a GIS, two- or three-dimensional analytic geometry is utilized

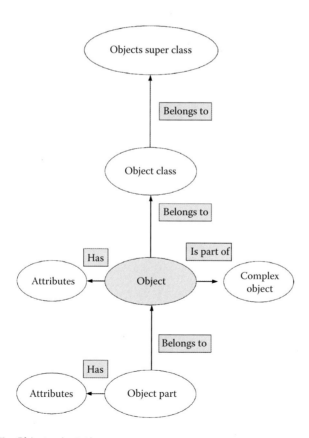

Figure 4.15 Object orientation.

for the formulations. In two dimensions, the following terms may be calculated from coordinates:

- Distances:

$$d_{ij} = \sqrt{(x_i - x_j)^2 + (y_i - y_j)^2}$$

- Directions (for a system in which x points up and y to the right):

$$\alpha_{ij} = arctg \frac{y_i - y_j}{x_i - x_j}$$

- New point positions:

$$x_i = x_i + d_{ij} \sin \alpha_{ij}$$
$$y_j = y_i + d_{ij} \cos \alpha_{ij}$$

- Angles:

$$\beta_{ijk} = \alpha_{ik} - \alpha_{ij}$$

- Areas of a polygon:

$$A = \frac{1}{2} \sum_{i=1}^{n} (x_i + x_{i+1})(y_{i+1} - y_i)$$

- Center of mass of a polygon:

$$x_c = \frac{1}{6A} \sum_{i=1}^{n} (x_i + x_{i+1})(x_i y_{i+1} - y_i x_{i+1})$$

$$y_c = \frac{1}{6A} \sum_{i=1}^{n} (y_i + y_{i+1})(x_i y_{i+1} - y_i x_{i+1})$$

- Circumference of a polygon:

$$C = \sum_{i=1}^{n} \sqrt{(x_i - x_{i+1})^2 (y_i - y_{i+1})^2}$$

For three dimensions, these expressions are expandable as stated for solid analytical geometry.

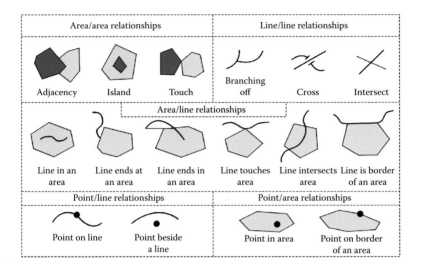

Figure 4.16   Geometric spatial queries.

Geometric spatial queries such as those shown in Figure 4.16 depend on calculations of intersections between points, lines, and areas.

The equation of a straight line in two dimensions is given by

$$Ax_i + By_i + C = 0$$

Using the coordinates of the two endpoints defining the line, the formulation is:

$$\frac{y_i - y_1}{y_2 - y_1} = \frac{x_i - x_1}{x_2 - x_1}$$

or

$$\begin{vmatrix} x & y & 1 \\ x_1 & y_1 & 1 \\ x_2 & y_2 & 1 \end{vmatrix} = 0$$

The shortest distance, d, of a point, P, from a straight line becomes

$$d = \frac{Ax_p + By_p + C}{\sqrt{A^2 + B^2}}$$

Two lines

$$A_1 x_i + B_1 y_i + C_1 = 0$$
$$A_2 x_i + B_2 y_i + C_2 = 0$$

may be intersected. The point of intersection has the coordinates

$$x_i = \frac{B_1 C_2 - B_2 C_1}{A_1 B_2 - A_2 B_1}$$

$$y_i = \frac{C_1 A_2 - C_2 A_1}{A_1 B_2 - A_2 B_1}$$

The straight lines are parallel if $A_1 B_2 - A_2 B_1 = 0$. They are perpendicular if $A_1 A_2 + B_1 B_2 = 0$.

The distance, d, between the parallel straight lines is:

$$d = \frac{C_1 - C_2}{\sqrt{A_1^2 + B_1^2}}$$

The equation of a circle is:

$$x_i^2 + y_i^2 + A x_i + B y_i + C = 0$$

or

$$(x_i - x_o)^2 + (y_i - y_o)^2 = r^2$$

Using the tools of analytic geometry, these formulations may be expanded to second and higher-order curves.

In three dimensions, the equivalent terms become:

- Distances:

$$d_{ij} = \sqrt{(x_i - x_j)^2 + (y_i - y_j)^2 + (z_i - z_j)^2}$$

- Direction cosines:

$$\cos \alpha_x = \frac{(x_i - x_j)}{d_{ij}}$$

$$\cos \alpha_y = \frac{(y_i - y_j)}{d_{ij}}$$

$$\cos \alpha_z = \frac{(z_i - z_j)}{d_{ij}}$$

- The spatial angle between directions with the direction cosines α_x, α_y, α_z, and α'_x, α'_y, α'_z is:

$$\cos \beta = \cos \alpha_x \cos\alpha'_x + \cos\alpha_y \cos\alpha'_y + \cos \alpha_z \cos\alpha'_z$$

The equation of a plane determined by three points becomes:

$$Ax_i + By_i + Cz_1 + D = 0$$

or

$$\begin{vmatrix} x_i & y_i & z_i & 1 \\ x_1 & y_1 & z_1 & 1 \\ x_2 & y_2 & z_2 & 1 \\ x_3 & y_3 & z_3 & 1 \end{vmatrix} = 0$$

A straight line in three dimensions is defined by two independent linear equations:

$$A_1x_i + B_1y_i + C_1z_1 + D_1 = 0$$
$$A_2x_i + B_2y_i + C_2z_1 + D_2 = 0$$

or

$$\frac{x_i - x_1}{x_2 - x_1} = \frac{y_i - y_1}{y_2 - y_1} = \frac{z_i - z_1}{z_2 - z_1}$$

A distance, d, between a point, P, and a plane is given by:

$$d = \frac{Ax_p + By_p + C_p + D}{\sqrt{A^2 + B^2 + C^2}}$$

So far, most GIS systems have been limited to two-dimensional geometry. Elevations have, however, been included as attributes. One therefore speaks of a two-and-a-half-dimensional capability. The elevation information may be introduced on a case-by-case basis, if a special three-dimensional query is desired and programmed.

With the described tools of analytic geometry, the GIS vector application software tasks listed in this chapter (see the earlier section "Vector Geometry") may be programmed, for example:

- *Translations*, Δx and Δy, are possible by:

$$\begin{pmatrix} x' \\ y' \end{pmatrix} = \begin{pmatrix} x + \Delta x \\ y + \Delta y \end{pmatrix}$$

- Rotations by an angle, α, are determined by:

$$\begin{pmatrix} x' \\ y' \end{pmatrix} = \begin{pmatrix} \cos\alpha - \sin\alpha \\ \sin\alpha\cos\alpha \end{pmatrix} \begin{pmatrix} x \\ y \end{pmatrix}$$

- *Scale change* by a scale factor, λ, is executed by:

$$\begin{pmatrix} x' \\ y' \end{pmatrix} = \lambda \begin{pmatrix} x \\ y \end{pmatrix}$$

- *Edge cut-off* is calculated by the calculation of intermediate points intersecting the equation of a line with the boundaries of a rectangle, $x_{max}y_{max}x_{min}y_{min}$, of the area to be displayed.

RASTER SYSTEMS

Raster data consist of a regular 2D grid of square cells. The grid is characterized by a (geocoded) origin, its (geocoded) orientation, and the raster cell size, which for imagery corresponds to a pixel (picture element) size. Other information, such as elevation levels or thematic data, may also be arranged by a scheme of regular tessellation. Raster systems may also be arranged in three dimensions. The 3D cell becomes a cube (a voxel). The attribute of the cell describes the thematic information (gray level, elevation level, thematic object content). Raster coverages may be in regular (square, rectangular) or irregular dimensions. Each raster data set constitutes a layer. There may be many layers for the same area.

The geometric accuracy of raster data is limited by the cell resolution. A mixed-cell problem may exist. Due to limits in resolution, there is a possibility of mixed pixels.

Whereas raster models reflect what is present, vector models more accurately define the whereabouts. Raster topology is defined by the eight neighboring pixels surrounding a particular pixel. Neighboring cells carrying the same attribute define a connection component. In this way linear objects may be recognized by the connection components.

Raster data operations are possible in the following ways:

- *Geometric transformations*, which permit geocoding via digital (differential) rectification and the presentation of perspective views including its geometric resampling algorithms.
- *Radiometric transformations*, which include all types of digital image processing (filtering, multispectral classification, image analysis).

Geometric and radiometric transformations have already been discussed in previous chapters.

- *Algebraic transformations*, which permit the combination of different layers and their analysis by Boolean operators. Boolean operators are AND, OR, XOR, and NOT. AND signifies if both layer pixels have a particular value. OR signifies if one of them has that particular value. XOR means the exclusive OR, that both do not have that particular value, and NOT signifies that one of them does not have that particular value.
- *Macro-operations*, applied for line enhancement or line thinning, are the Blow and Shrink operations. They consist of a shift of the raster image in all four directions and the logical OR application. This thickens the line. For line thinning the same operation is applied for the background. Holes that may have resulted from the operation are filled by filtering.

DATABASES

A database is a self-contained, long-term organization of data for flexible and secure use. It consists of the data and of a database management system, the software to manage the data. A database permits a strict separation between data and an application. It has a well-defined interface for application programs. The user of a database is not concerned with the internal data organization, but he or she can change the data location without changing the application program. The database management system provides efficient access to the data with security checks.

The internal view of a database is the physical memory allocation. The conceptual view concerns the logical data organization, and the external view to the user is the graphic user interface. Databases are used in a large variety of applications (banking, reservation systems, libraries, business). GIS can take advantage of these developments.

Basic to the generation of a database is the entity–relationship model for a particular application. The relationship defines the association between entity types, for example:

Point number–coordinates, 1:2
Building–parcel, n:1
Parcel–village, m:1

Metadata are stored together with the data. For example, they define the reference system, the resolution, and the date of data acquisition.

Database structures may be

- Hierarchical
- Network
- Relational
- Object–relational

Hierarchical databases are the oldest database type. They support 1:*n* relationships well but lead to redundant storage for *n:m* relationships. They are based on a tree structure, and they are inflexible with regard to changes.

Network databases have been available since the 1970s, when the Conference on Data System Languages (CODASYL) introduced the first commercially available database of this type. It supports 1:*m* and *n:m* relationships without redundancy. It was used in older GIS systems.

The relational database is currently the most common database model. It is based on relational tables. Each entity is characterized by a table, called a "tuple." Relationships between different entity types are also expressed by tables. The sequence of tuples is irrelevant. Unique access to an individual entity is made possible by a segment having a 1:1 or *n*:1 relation to other segments (e.g., point number, position). The organization of each table is independent of other tables. Tables can be accessed, combined, and changed by simple operations. Masking permits different external views of the data. Simple rules prevent redundancy. The disadvantage of relational databases is that they are designed for simple data. For 2D topological and for 3D spatial queries they become slow for interactive work. Thus, at least temporary extensions in GIS systems can accelerate the management of interactive sessions. Microsoft's Data Access is a typical relational database.

Object–relational databases are extensions of relational databases toward object orientation. They contain user-defined data types as domains of a table and user-defined functions. Examples for object–relational databases are

- Oracle 8, Spatial Data Cartridge
- Esri, Spatial Database Engine SDE
- Informix, 2D-Spatial Data blade

Many public data collections have a spatial reference in the form of a location name, called "indirect spatial access," for example, street address and city.

GIS data have a "direct spatial reference" via 2D or 3D coordinates. To use indirect spatial access, data in GIS geocoded links for these must first be established (e.g., by a parcel number, a building number, or a set of coordinates). In very large databases, a coordinate reference is usually added to the data to allow for efficient retrieval.

The geometric and thematic data of a certain area are stored together in regular raster cell divisions of the database. This map-based (tile-based) organization ensures quick access to the data (ArcInfo).

Another possibility is to organize the raster cells in the form of a quad tree structure. The objective is to store approximately equal amounts of data content in each cell. If a regular raster cell in a coarse raster cell division contains more data than a limit allows, then the raster cell is subdivided consecutively in half or quarter dimension cells. Spatial indexing is used for quick access of the data sets (SICAD, Oracle Spatial). Other cell subdivisions, such as the k-dimensional tree (K-D tree) or the grid file method with irregular raster cell dimensions, are possible. The R-Tree (Intergraph-Tigris) utilizes a minimum-bounding rectangle for entities and groups of entities determined by clustering.

GIS SYSTEMS

There are a few hundred GIS systems in existence. Web sites of some of the major international GIS vendors are:

- Bentley (United States): www.bentley.com
- Caris (Canada): www.caris.com
- Erdas (United States): www.geosystems.de
- Esri (United States): www.esri.com
- Genasys (Australia): www.genasys.com
- GE-Smallworld (United Kingdom): www.gedistalenergy.com/GIS.htm
- Idrisi (United States): www.clarklabs.org
- Intergraph (United States): www.intergraph.com
- MapInfo (United States): www.mapinfo.com
- PCI Geomatics (Canada): www.pcigeomatics.com
- SICAD (Germany): www.aed-sicad.de

On these and more recent Web sites, the various software components offered by these companies can be traced.

The GIS vendors have realized the weakness of producing their data in proprietary formats. Raster systems TIFF (Tagged Image File Format) and Geo TIFF format (with metadata for georeferencing) presented few difficulties. These formats may be compressed into the JPEG (Joint Photographic Experts Group) format at variable compression rates.

The vector data format conversion, for example, between Esri formats (dxf and shp files), Intergraph formats (dgn files), SICAD formats (EDBS files), or AutoCAD formats (dwg files) necessitated more or less effective conversion programs. With the Open-GIS Consortium, vendors of GIS systems cooperate in attempts to ensure transferability of data generated by the particular vendor system into that of another vendor. Various working groups of the International Standards Organization (ISO), in particular of the Technical Committee TC 211 on Geographic Information and Geomatics, support this effort.

GIS and the Internet

Most GIS vendors in their early development provided access to an Intranet or the Internet, for example:

- Internet Map Server and Map Objects by Esri
- Geomedia Web Map by Intergraph
- Smallworld Web by Smallworld

The Intranet and Internet do not differ in technology but they regulate access. The Intranet is limited to use by a company or an administration, whereas the Internet is accessible, for example, through the World Wide Web, often at a cost.

The data reside in geodata servers and are remotely accessed. The client in the Web accesses the server via plug-ins or applets. These are small programs, which can be loaded from the Internet. Java applets are independent of a specific platform. The server contains a common gateway interface. The client makes a request at the Web or FTP server through the Web browser. The request is passed on to the common gateway interface to start the application (e.g., to retrieve a particular data set). The answer is passed on through the gateway interface and the Web or the FTP server to the client. FTP servers permit the exchange of large data sets. The connection between client and server may be through wired telephone connections, which depending on the particular service, can vary in performance from 19 kbaud (19kb/sec) to 64 kbaud for ISDN connections, to several megabauds for Ethernet or glass fiber connections.

But now wireless connections are offered by mobile telephone companies. This is in line with the development of mobile communication networks; for example, the European GSM system was replaced by the UMTS mobile phone system, permitting higher transmission rates. The transmission of data for wireless connections is regulated through the wireless application protocol (WAP). The mobile phone industry and the industry producing personal digital assistants (PDA), such as Hewlett-Packard, Nokia, Palm, Psion, Apple, and Samsung, have provided opportunities for mobile GIS users through smartphones and tablets.

ESRI's ArcGIS

The company Environmental Systems Research Institute (Esri) was established in 1969 by Jack Dangermond in Redlands, California. It is globally the major producer and vendor of GIS systems. For this reason, special attention has been given in this book to describe ArcGIS with its components as an example for the state of the art, which continues to develop from one program version to the next.

Esri developed the first modern GIS on minicomputers under the name ArcInfo, which was released in 1982. ArcInfo was originally Fortran based. Later, a separate command language, AML, was used. With ArcInfo version 5 and with version 7.1 it became accessible to Sun Solaris and to Windows operating systems on the desktop.

In 1999, ArcGIS 8.0 became the new desktop platform. It was introduced at four levels (Figure 4.17).

- ArcReader; it is free of charge and only permits map viewing

The three other versions with increasing capabilities were licensed:

- ArcView for map viewing and data query
- ArcEditor for mapping, geocoding, editing, and customized operation
- ArcInfo for advanced editing and advanced geoprocessing

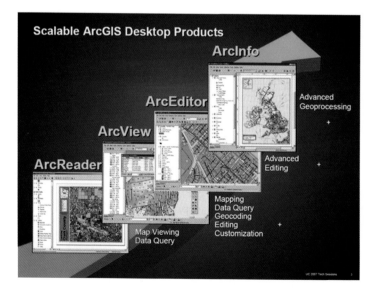

Figure 4.17   The four levels of ArcGIS. (From Esri, Redlands, California.)

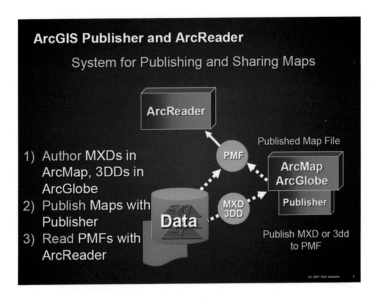

Figure 4.18   ArcReader and ArcPublisher. (From Esri, Redlands, California.)

ArcReader and ArcPublisher are program components for viewing and publishing map data in ArcMap in 2D or in ArcGlobe form in 3D (Figure 4.18). ArcReader permits display functions, such as finding 2D maps, panning and zooming them (Figure 4.19), or viewing or flying over 3D ArcGlobe scenes (Figure 4.20).

ArcPublisher creates the settings for each map to be viewed or published (see Figure 4.21). To create or edit the content of each map, tools are provided to digitize continuous lines, or link point-by-point geometry as a digitizing process, or to enter or derive coordinate values by COGO; to create areas; and to add attributes to points, lines or areas (see Figure 4.22). To do this, ArcMap has the tools to edit via a toolbar in which the different editing options are selected (see Figure 4.23). For this workflows are created (Figure 4.24), in which action options are selected for line snapping (Figure 4.25).

To maintain geometric analysis capabilities, 2D topology information is maintained using shapefiles. For features defined by the shapefiles, such as points, lines, and areas, attributes may be created and updated. Attributes may be edited (Figure 4.26).

Other tools exist for changing the geometry of area features (Figure 4.27) or to create parallel or perpendicular line features at a certain distance (Figure 4.28).

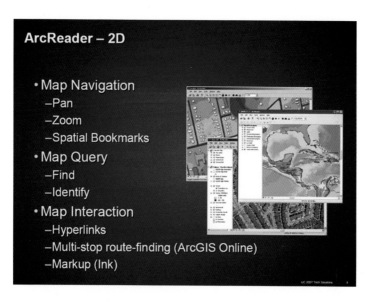

Figure 4.19 ArcReader as 2D viewer. (From Esri, Redlands, California.)

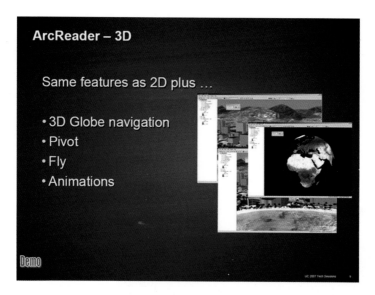

Figure 4.20 ArcReader as ArcGlobe viewer. (From Esri, Redlands, California.)

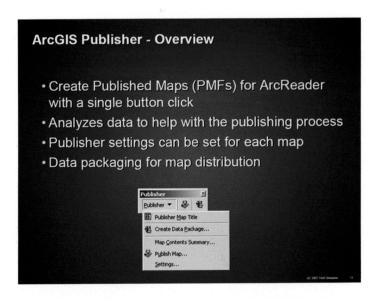

Figure 4.21   ArcPublisher settings. (From Esri, Redlands, California.)

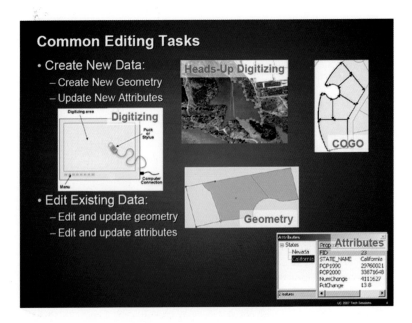

Figure 4.22   Geometry edit options. (From Esri, Redlands, California.)

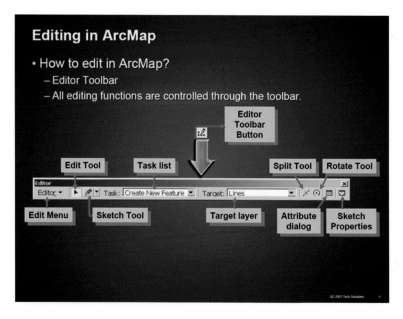

Figure 4.23   ArcMap edit with toolbar. (From Esri, Redlands, California.)

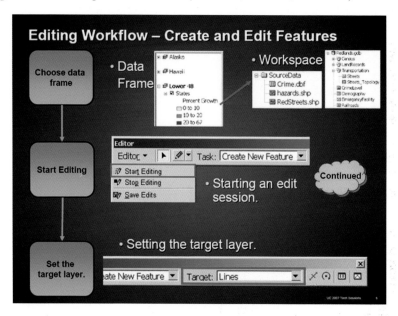

Figure 4.24   Editing workflow. (From Esri, Redlands, California.)

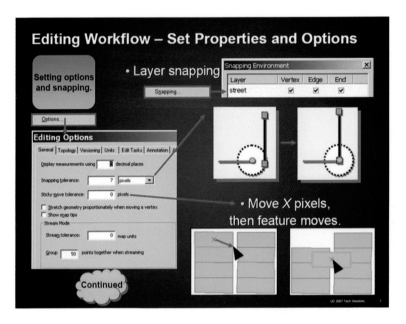

Figure 4.25   Line snapping. (From Esri, Redlands, California.)

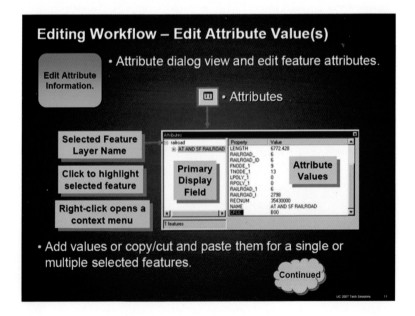

Figure 4.26   Editing of attributes. (From Esri, Redlands, California.)

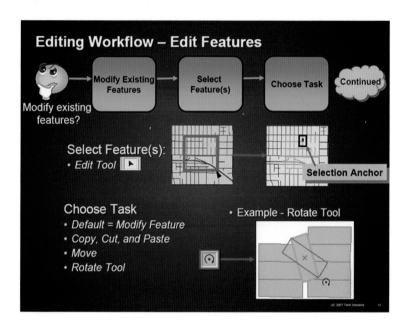

Figure 4.27   Geometry changes for entire areas. (From Esri, Redlands, California.)

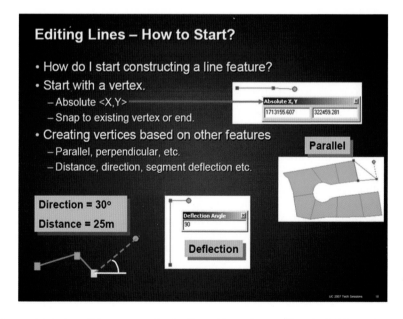

Figure 4.28   Parallel or perpendicular lines. (From Esri, Redlands, California.)

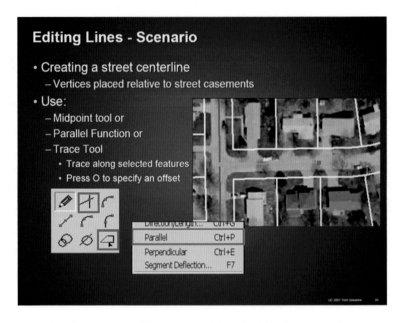

Figure 4.29   Midpoint or midline creation. (From Esri, Redlands, California.)

Other tools are for the creation of midpoints or street center lines (Figure 4.29) or for special features, such as cul-de-sacs (Figure 4.30). Or a line may be partitioned at a certain interval (Figure 4.31). Special tools are provided to create polygons independently (Figure 4.32) or adjacent to the polygons (Figure 4.33). Polygons may of course also be created with COGO (Figure 4.34). Other possibilities exist for the merger of polygons (Figure 4.35) or to intersect polygons (Figure 4.36).

The data contained in a database are administered through ArcCatalog (Figure 4.37). They are shown in a catalog tree. A right click on the mouse permits one to select the task to do with the data file (copy, export, delete). ArcCatalog permits one to search and display the data in its various forms (images, maps) (Figure 4.38). The catalog tree identifies the various data types by icons (Figure 4.39). ArcCatalog permits one to manage the data by creating geodatabases, by defining the structure in ArcCatalog and by editing the content in ArcMap (Figure 4.40). For the catalog tree, feature classes may be defined as well as their geometric reference (Figure 4.41). Or the formats of the data may be converted for inputs into the geodatabase (Figure 4.42). Multiple data source types may be merged (shapefiles, CAD formats) (Figure 4.43). Also, different raster formats may be used (jpeg, tiff) (Figure 4.44). ArcCatalog also permits graphic searches (Figure 4.45).

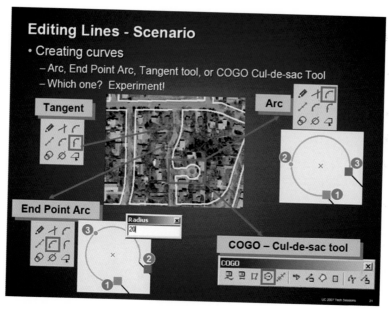

Figure 4.30   Special geometry features. (From Esri, Redlands, California.)

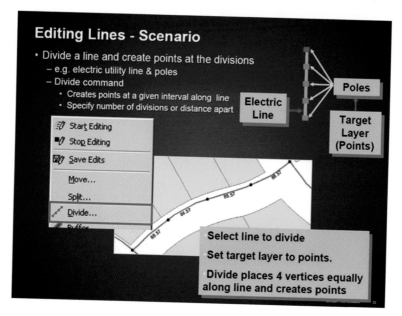

Figure 4.31   Partitioning lines. (From Esri, Redlands, California.)

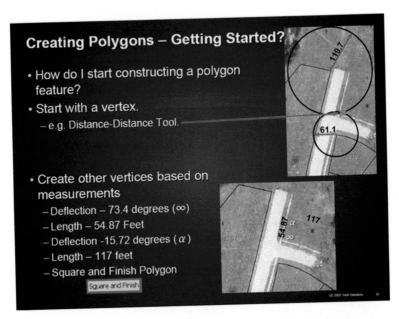

Figure 4.32   Independent polygon creation. (From Esri, Redlands, California.)

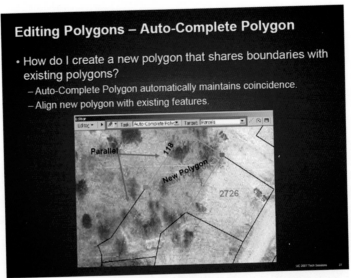

Figure 4.33   Completion of polygons. (From Esri, Redlands, California.)

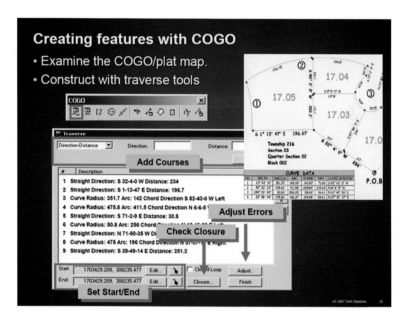

Figure 4.34   Feature creation with COGO. (From Esri, Redlands, California.)

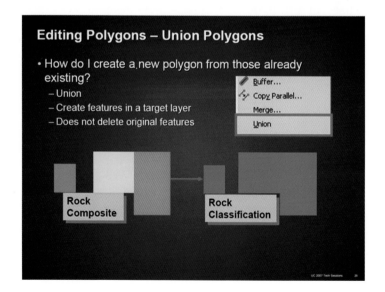

Figure 4.35   Merger of polygons. (From Esri, Redlands, California.)

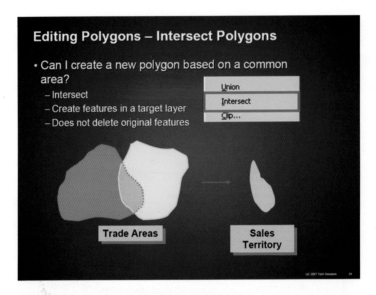

Figure 4.36   Intersection of polygons. (From Esri, Redlands, California.)

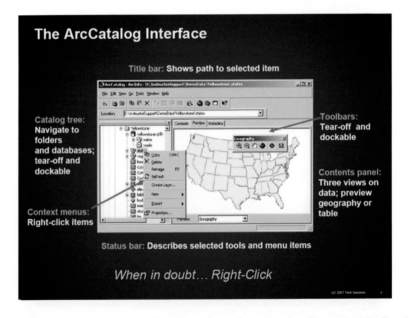

Figure 4.37   Administration of database through ArcCatalog. (From Esri, Redlands, California.)

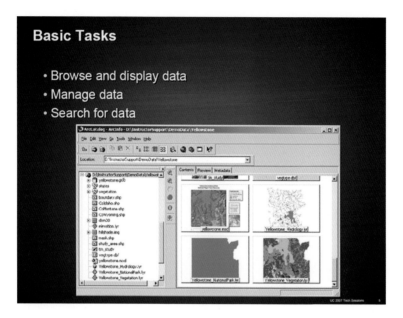

Figure 4.38   Search of data. (From Esri, Redlands, California.)

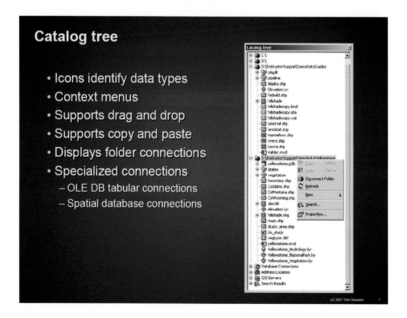

Figure 4.39   Use of catalog tree. (From Esri, Redlands, California.)

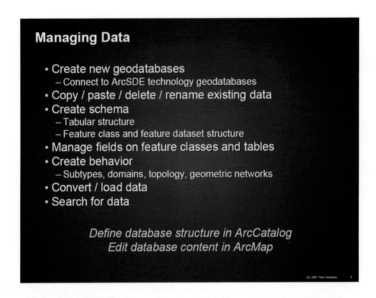

Figure 4.40 Management of data by ArcCatalog. (From Esri, Redlands, California.)

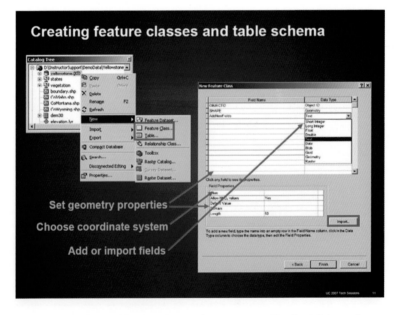

Figure 4.41 Definition of feature classes. (From Esri, Redlands, California.)

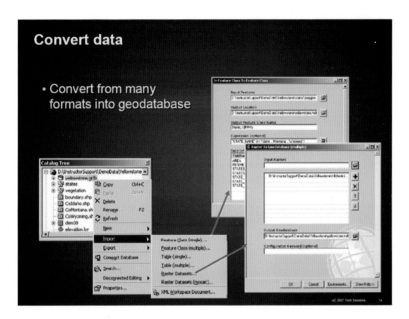

Figure 4.42   Format conversions. (From Esri, Redlands, California.)

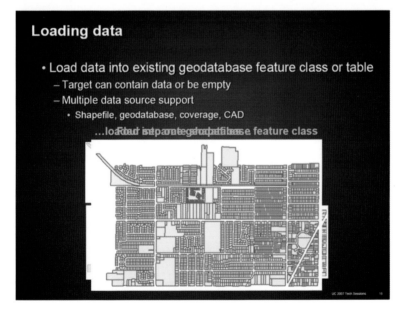

Figure 4.43   Loading of vector data in various formats. (From Esri, Redlands, California.)

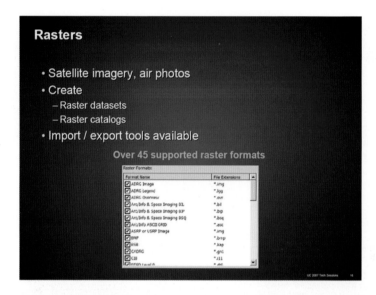

Figure 4.44   Use of different raster formats. (From Esri, Redlands, California.)

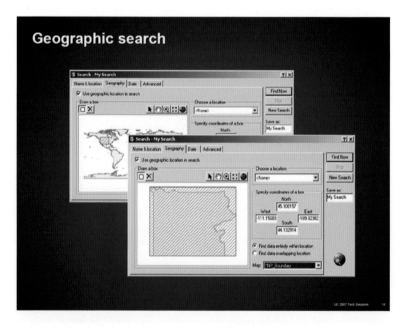

Figure 4.45   Graphical searches. (From Esri, Redlands, California.)

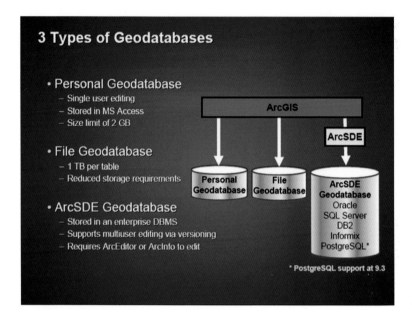

Figure 4.46   Types of geodatabases. (From Esri, Redlands, California.)

Depending on the volume of data the following types of geodatabases may be used in ArcGIS (Figure 4.46):

- For a data size <2GB, a Personal Geodatabase in Microsoft Access
- For a data size <1TB, a File Geodatabase
- For an unlimited data size, an ArcSDE-based geodatabase, which may be connected to other large relational databases of other vendors, such as Oracle, SQL Server, DB2, Informix, or PostgreSQL. The characteristics of these types are described in Figure 4.47. A summary of the management possibilities for geodatabases is shown in Figure 4.48.

In addition, it is possible to describe data sets by metadata with respect to (see Figure 4.49):

- Content
- Quality
- Condition
- Origin

The strength of a GIS System, such as ArcGIS, are of course the various analysis capabilities. For all licensing levels for ArcGIS, special purpose extensions for different types of spatial analysis are available:

3 Types of Geodatabases...

	Personal GDB	File GDB	SDE GDB (3 editions)
Storage format	Microsoft Access	Folder of binary files	DBMS
Storage capacity	2 GB	1 TB per table*	Depends on edition
Supported O/S platform	Windows	Any platform	Depends on edition
Number of users	Single editor Multiple readers	Single editor Multiple readers	Multiple editors & readers
Distributed GDB functionality	Check out/check in replication	Check out/check in replication	Replication (all types) & versioning

* By default; option to have 256 TB per table

Figure 4.47   Properties of geodatabase types. (From Esri, Redlands, California.)

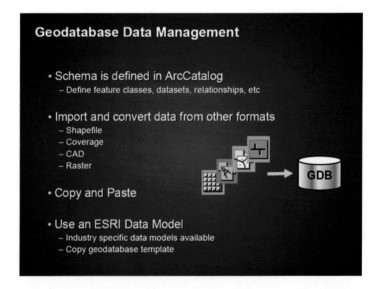

Figure 4.48   Geodatabase data management. (From Esri, Redlands, California.)

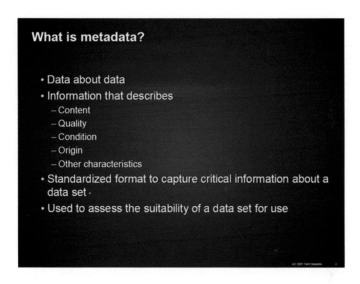

Figure 4.49   Metadata. (From Esri, Redlands, California.)

- The spatial analyst is an integrated raster and vector spatial analysis tool (Figure 4.50). The analysis tools available are listed in Figure 4.51. Of special interest is the map query with Boolean operations, shown in Figure 4.52. Other tools are for distance and proximity analysis (Figure 4.53) or for corridor analysis by cost models (Figure 4.54).

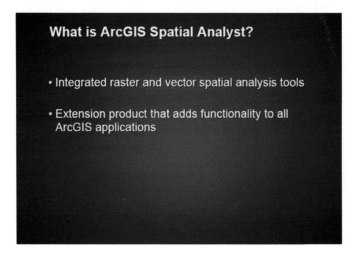

Figure 4.50   ArcGIS Spatial Analyst. (From Esri, Redlands, California.)

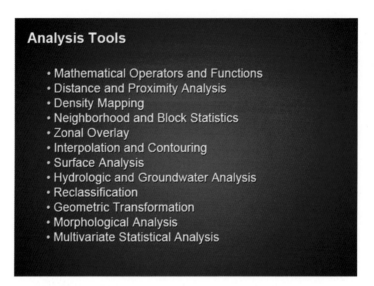

Figure 4.51   Spatial Analyst tools. (From Esri, Redlands, California.)

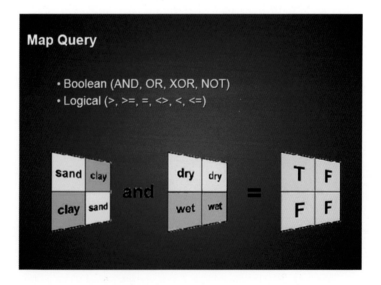

Figure 4.52   Boolean map query. (From Esri, Redlands, California.)

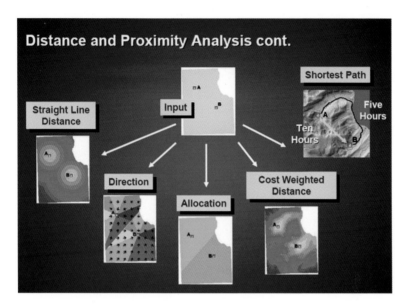

Figure 4.53   Proximity analysis. (From Esri, Redlands, California.)

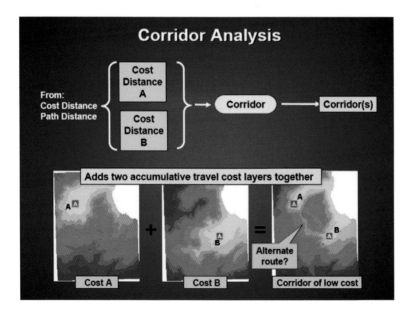

Figure 4.54   Corridor analysis. (From Esri, Redlands, California.)

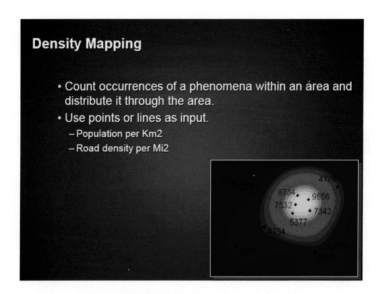

Figure 4.55   Density mapping. (From Esri, Redlands, California.)

Other tools are for density mapping by interpolation between points (Figure 4.55). Neighborhood properties may be analyzed by filters in raster data (Figure 4.56) or by zonal overlays for vector and raster data (Figure 4.57). Similar tools are available for contour interpolation from point data (see Figure 4.58) or for surface analysis using slope and aspect data from digital elevation models (DEMs) (see Figure 4.59). Solar radiation may also be modeled (Figure 4.60). Raster DEMs and vector watersheds and stream lines may also be used to perform a hydrological analysis (Figure 4.61). For multispectral images a maximum likelihood classification may be applied (Figure 4.62). Also, image processing filters may be applied to generalize raster data (Figure 4.63).

- The 3D analyst is an extension for 3D interactive visualization (Figure 4.64). It can be applied to 3D surface models (raster or triangulated irregular networks [TINs]) and to 3D vectors (Figure 4.65). It can be applied within ArcGlobe, ArcScene, ArcMap, and ArcCatalog (Figure 4.66). It is suitable to create animations and to export them as videos (Figure 4.67). It is addressable by a toolbar (Figure 4.68). ArcScene has many added possibilities for us (Figure 4.69).
- The network analyst is an extension for analysis in transportation networks (see Figure 4.70). Due to its real-time uses in the field with mobile devices, it is closely linked to ArcGIS Server, ArcGIS mobile,

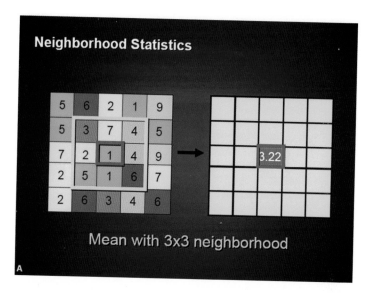

Figure 4.56   Raster neighborhood statistics. (From Esri, Redlands, California.)

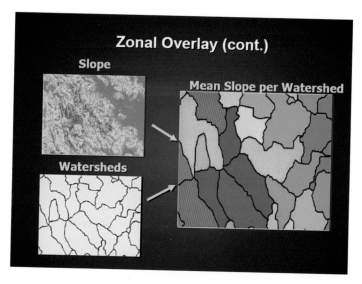

Figure 4.57   Zonal overlays. (From Esri, Redlands, California.)

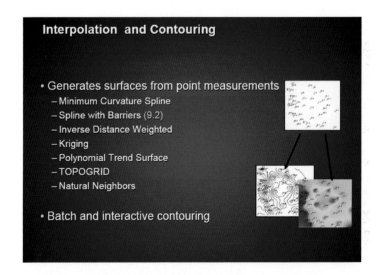

Figure 4.58   Interpolation and contouring. (From Esri, Redlands, California.)

ArcIMS, and ArcPad (Figure 4.71). Its route solver may find the closest or fastest route between two road network points (Figure 4.72). Or it may find the closest facility within the network (Figure 4.73). The service Area Solver may find the area that can be traversed at a specified cost (Figure 4.74). For multiple route options it is able to derive cost

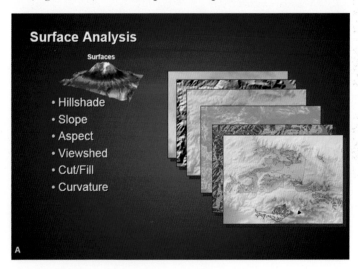

Figure 4.59   Surface analysis. (From Esri, Redlands, California.)

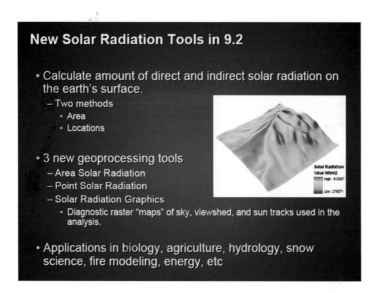

Figure 4.60   Solar illumination. (From Esri, Redlands, California.)

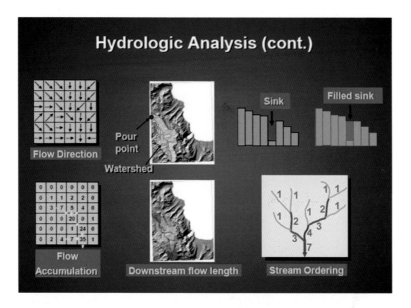

Figure 4.61   Hydrological analysis. (From Esri, Redlands, California.)

Figure 4.62   Multispectral classification tools. (From Esri, Redlands, California.)

models for these (Figure 4.75). Solver options with barriers or restrictions may be included (Figure 4.76). A hierarchy between higher- and lower-order routes may also be considered (Figure 4.77). To each route section, attributes for calculating cost models may be added (Figure 4.78).

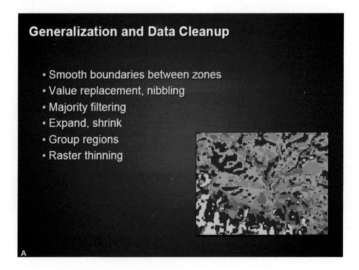

Figure 4.63   Raster data cleanup by generalization. (From Esri, Redlands, California.)

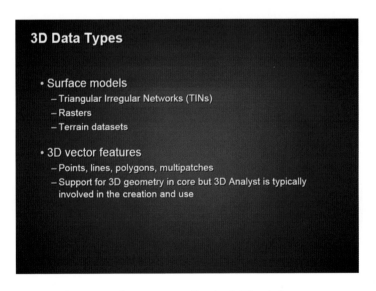

What is 3D Analyst?

- An extension that adds capabilities for
  - Interactive 3D visualization of spatial data
  - Surface creation & analysis

- Provides 3D GIS capabilities using
  - Stand-alone visualization applications: ArcGlobe & ArcScene
  - 3D Geoprocessing tools
  - Interactive exploratory tools in ArcMap
  - 3D previewing and manipulation in ArcCatalog
  - Publish globe services (ArcGIS Server) for use in ArcGlobe
  - Publish globe documents (Publisher toolbar) for use in ArcReader
  - TIN extension support in ArcInfo Workstation

Figure 4.64   3D Analyst. (From Esri, Redlands, California.)

3D Data Types

- Surface models
  - Triangular Irregular Networks (TINs)
  - Rasters
  - Terrain datasets

- 3D vector features
  - Points, lines, polygons, multipatches
  - Support for 3D geometry in core but 3D Analyst is typically involved in the creation and use

Figure 4.65   3D data types. (From Esri, Redlands, California.)

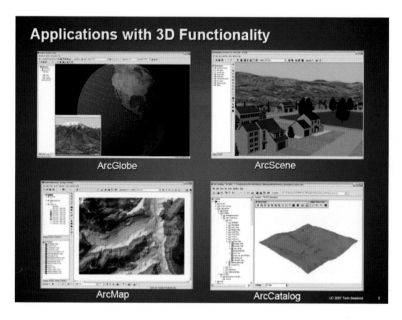

Figure 4.66   3D applications. (From Esri, Redlands, California.)

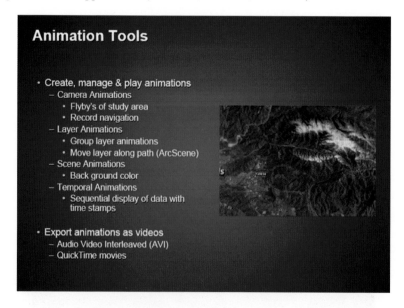

Figure 4.67   Animation tools of 3D Analyst. (From Esri, Redlands, California.)

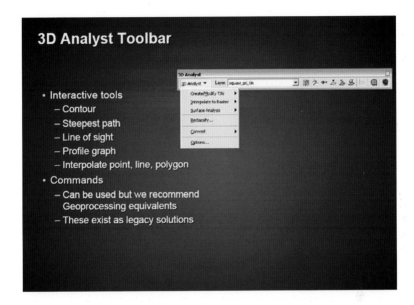

Figure 4.68   3D Analyst toolbar. (From Esri, Redlands, California.)

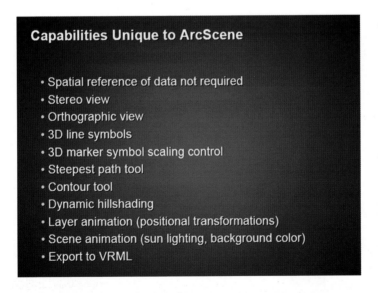

Figure 4.69   ArcScene capabilities. (From Esri, Redlands, California.)

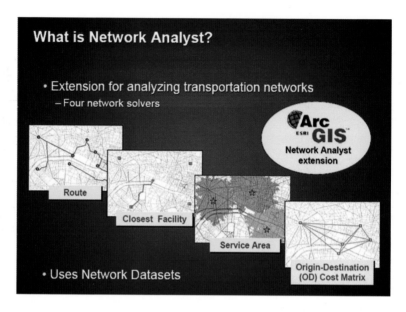

Figure 4.70   Network Analyst. (From Esri, Redlands, California.)

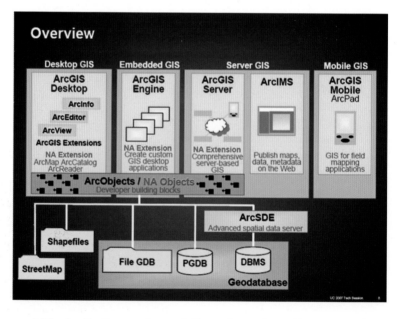

Figure 4.71   Application options of Network Analyst. (From Esri, Redlands, California.)

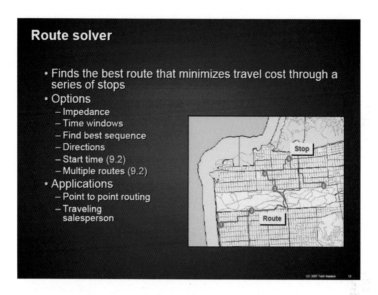

Figure 4.72   Route solver. (From Esri, Redlands, California.)

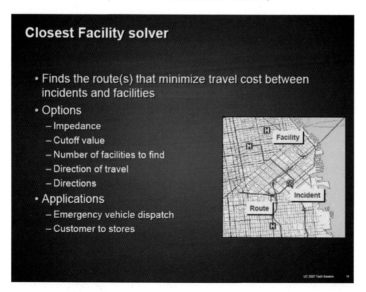

Figure 4.73   Closest facility solver. (From Esri, Redlands, California.)

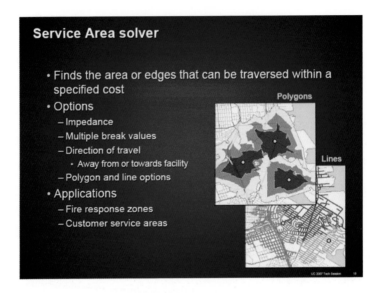

Figure 4.74   Service area solver. (From Esri, Redlands, California.)

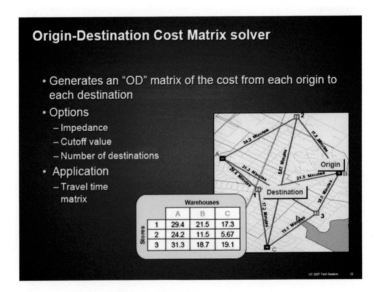

Figure 4.75   Cost matrix solver in network. (From Esri, Redlands, California.)

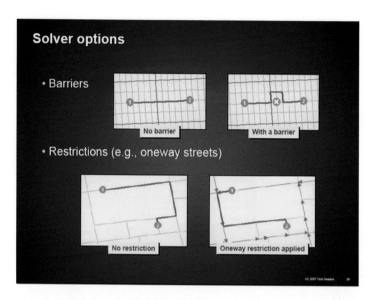

Figure 4.76   Options for restrictions in network. (From Esri, Redlands, California.)

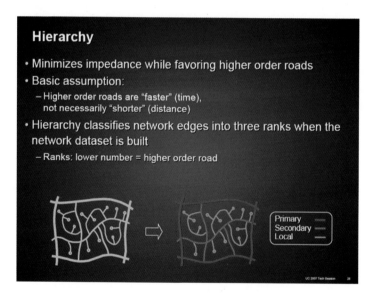

Figure 4.77   Hierarchies in networks. (From Esri, Redlands, California.)

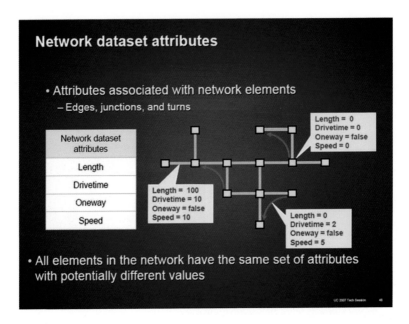

Figure 4.78 Attributes assigned to network sections. (From Esri, Redlands, California.)

- The statistical analyst, like the business analyst, is an extension to provide interpolation tools (Figure 4.79). The interpolation models may be deterministic (inverse distance weighting or polynomials), or they may be statistical (kriging) (Figure 4.80).

For all these tools geospatial processing programs are being used. To build an extension, by an Esri developer or by a customer, programming tools are provided in ArcGIS with ArcObjects (Figure 4.81).

Programming can be done with Visual Basic, C++, Java, or Python (Figure 4.82). They use files defined by ArcMap and ArcCatalog as objects (Figure 4.83), and they apply to all ArcGIS applications, desktop, the Web, or mobile (Figure 4.84). The Unified Modeling Language (UML) with its 70 libraries describes the ArcObjects classes (Figure 4.85).

A help to the ArcGIS user in setting up geoprocessing workflows and automating them is the Model Builder (Figure 4.86). The tools may be applied in a framework for automation (Figure 4.87). While specific tools may be applied by script or by command line, the Model Builder greatly simplifies the operation (Figure 4.88).

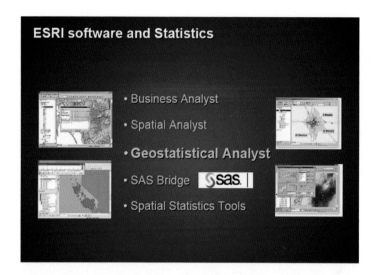

Figure 4.79   Geostatistical Analyst. (From Esri, Redlands, California.)

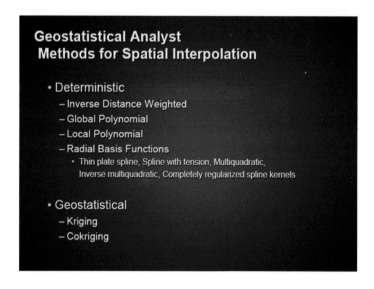

Figure 4.80   Use for interpolation. (From Esri, Redlands, California.)

Figure 4.81   Programming tools applied with ArcObjects. (From Esri, Redlands, California.)

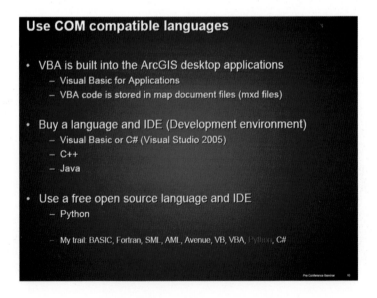

Figure 4.82   Programming language options. (From Esri, Redlands, California.)

Figure 4.83   Files defined as objects. (From Esri, Redlands, California.)

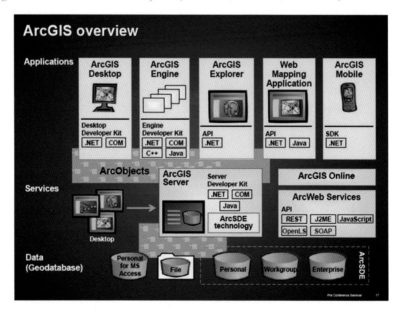

Figure 4.84   Objects defined for all ArcGIS applications. (From Esri, Redlands, California.)

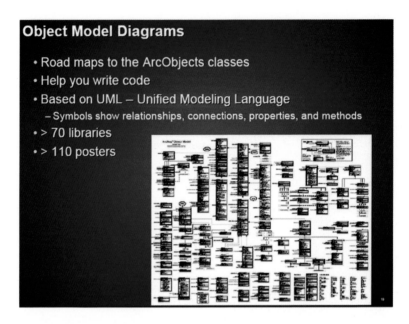

Figure 4.85 Objects defined as UML classes. (From Esri, Redlands, California.)

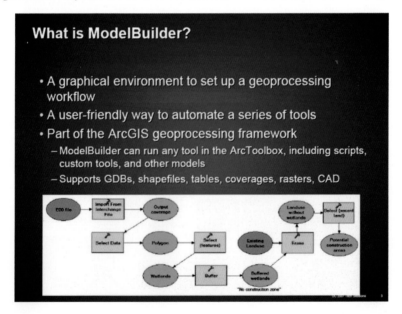

Figure 4.86 Workflows automated by Model Builder. (From Esri, Redlands, California.)

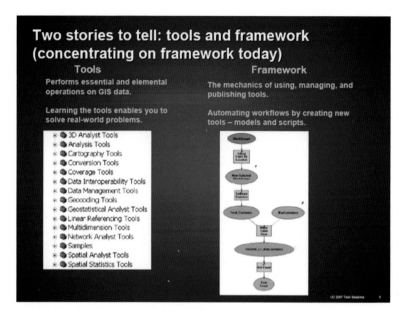

Figure 4.87   Framework for automation. (From Esri, Redlands, California.)

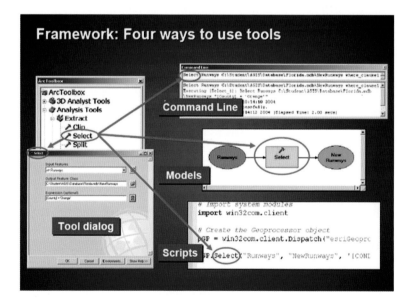

Figure 4.88   Options for generation of workflows. (From Esri, Redlands, California.)

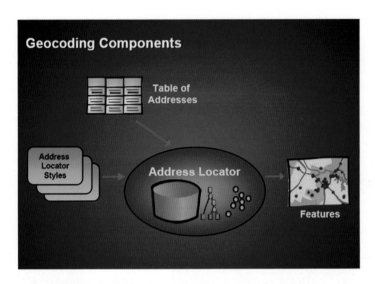

Figure 4.89   Address locator for geocoding. (From Esri, Redlands, California.)

With respect to geoinformation issues, topics of geocoding, cadastral data management, and photogrammetric operations, and the link to ground surveys and to the Web are of special interest. Here, ArcGIS offers a number of solutions:

1. The address locator, which links addresses to geocoded features (Figure 4.89).
2. The ArcGIS Cadastral Editor is used to generate a cadastral fabric and to administer land-related data (Figure 4.90). It is based on a topological data model (Figure 4.91). The parcel fabric can be built from survey data (Figure 4.92). Since such survey data are often not properly georeferenced, tools are provided to adjust survey data by least squares adjustment to a control network (Figure 4.93). In this way classical survey data may be fitted to more accurate GNSS (GPS) control (Figure 4.94).
3. The Image Server is an integrated photogrammetric and image processing tool within ArcGIS. It permits one to perform radiometric operations, such as pan sharpening or gray level stretching (Figure 4.95). It permits mosaicking of adjacent aerial images (Figures 4.96 and 4.97). It allows for rectification of images (Figure 4.98).
4. ArcPad, iPhones, or Android-based mobile phones are the link to ground surveys (Figure 4.99). The requirement is mobile wireless connectivity by telecommunication networks (Figure 4.100). These devices have many applications (Figures 4.101 and 4.102).

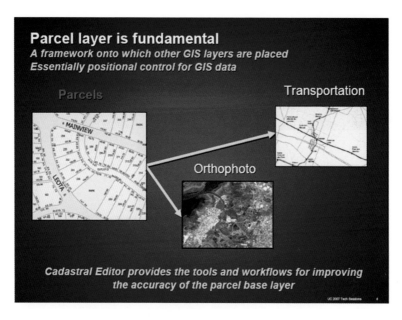

Figure 4.90   Cadastral Editor. (From Esri, Redlands, California.)

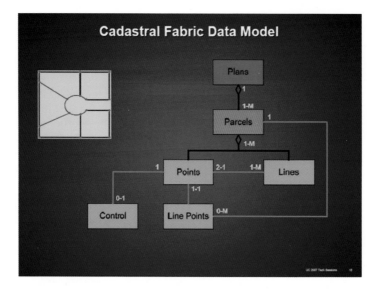

Figure 4.91   Topological data model for Cadastral Editor. (From Esri, Redlands, California.)

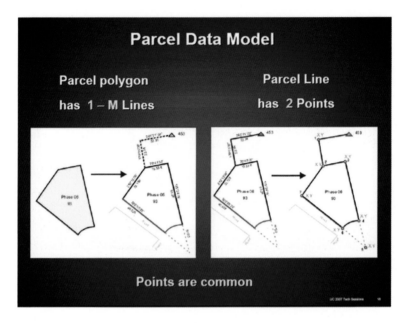

Figure 4.92 Building of cadastral fabric by survey data. (From Esri, Redlands, California.)

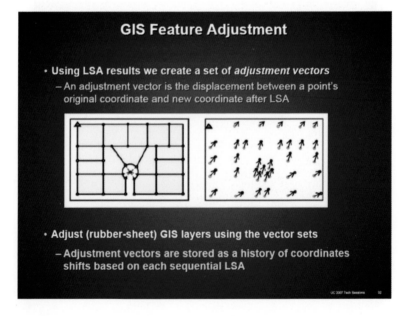

Figure 4.93 Adjustment of features by blocks. (From Esri, Redlands, California.)

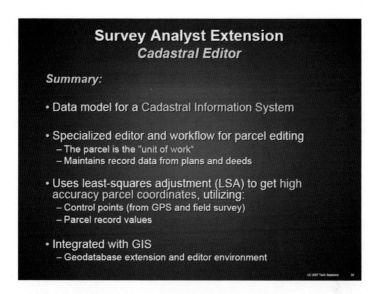

Figure 4.94   Use as cadastral information system. (From Esri, Redlands, California.)

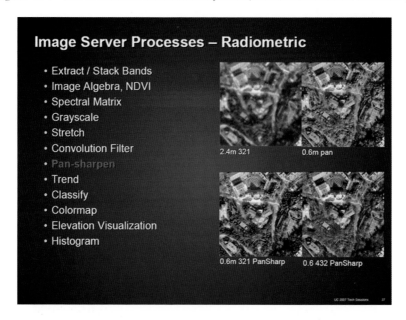

Figure 4.95   Image Server for radiometric changes of images. (From Esri, Redlands, California.)

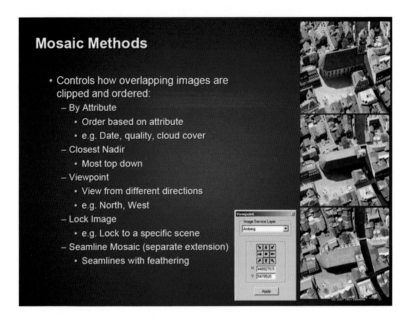

Figure 4.96   Fitting of adjacent imagery by Image Server. (From Esri, Redlands, California.)

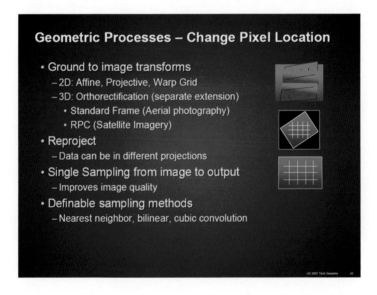

Figure 4.97   Geometric tools of Image Server. (From Esri, Redlands, California.)

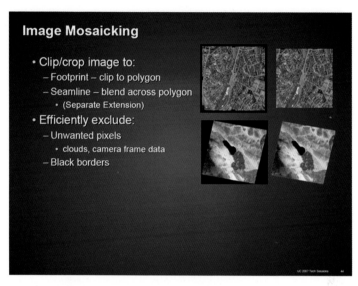

Figure 4.98   Image mosaicking by Image Server. (From Esri, Redlands, California.)

5. ArcWeb Services. Access to ArcGIS to the Web and to Web services is an important asset (Figure 4.103).
6. ArcGIS Enterprise is the future GIS trend based upon the established interface between ArcGIS and the Web on the way to a service-oriented infrastructure (Figure 4.104). Such an enterprise orientation

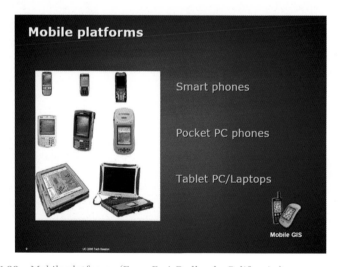

Figure 4.99   Mobile platforms. (From Esri, Redlands, California.)

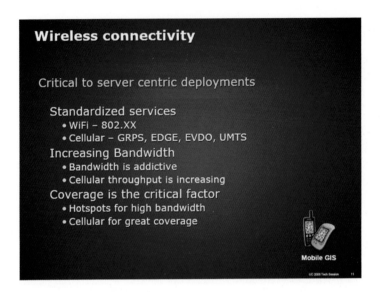

Figure 4.100   Wireless connectivity. (From Esri, Redlands, California.)

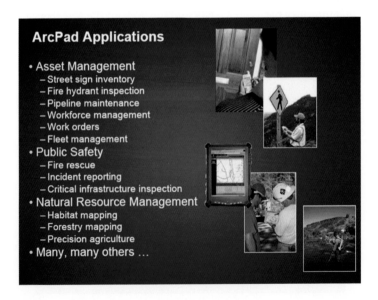

Figure 4.101   ArcPad applications. (From Esri, Redlands, California.)

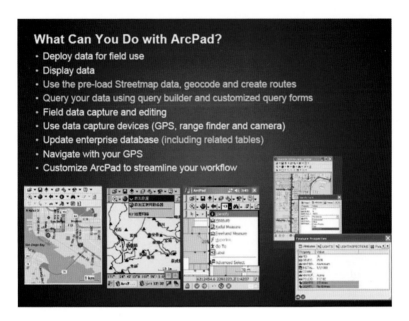

Figure 4.102   Mapping with ArcPad. (From Esri, Redlands, California.)

Figure 4.103   ArcWeb services. (From Esri, Redlands, California.)

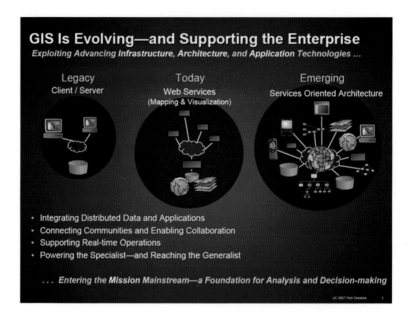

Figure 4.104   The evolving enterprise GIS. (From Esri, Redlands, California.)

is evolving in a number of global and crucial issues, such as geospatial intelligence, facilities management, land records administration,
business intelligence, and in the supply chain of goods (Figure 4.105).
To achieve an enterprise structure, data centralized GIS structures
are on the decline (Figure 4.106). Instead, distributed infrastructure is
on the rise (see Figure 4.107). Within such a structure, mobile devices,
servers, and enterprise systems become interconnected (Figure 4.108).
While GIS operations still continue in many ways, the tendency is
toward a service-oriented infrastructure (Figure 4.109).
7. ArcGIS Online. In the meantime, ArcGIS Online, available through
the Web at no charge, already interconnects different GIS components
through the Web (Figure 4.110).

During the Esri User Conference in San Diego in July 2013, version 10.2 of
ArcGIS was introduced. The emphasis was both on improvements to ArcGIS
Server for enterprise uses, as well as in parallel to ArcGIS Online, featuring the
use of Web GIS. In version 10.2 a portal is included. There is access to a rich data
set on authoritative base maps, and a recent and renewable coverage of imagery with 30 cm GSD over the United States, with 60 cm over Western Europe
and with 1 m GSD for remaining portions of interest of the world. This imagery
permits to update authoritative base maps for GIS users.

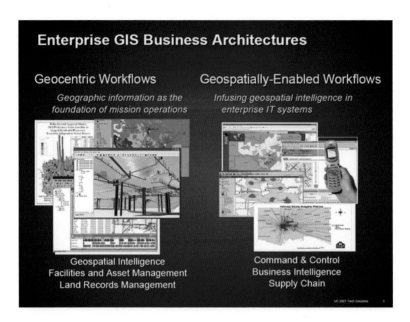

Figure 4.105   ArcGIS Enterprise applications. (From Esri, Redlands, California.)

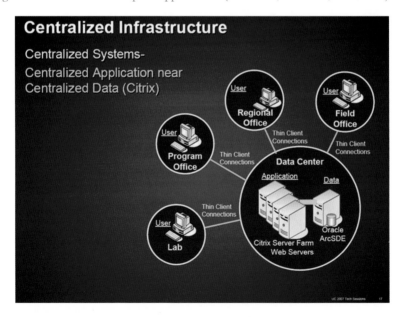

Figure 4.106   Centralized GIS structure of the past. (From Esri, Redlands, California.)

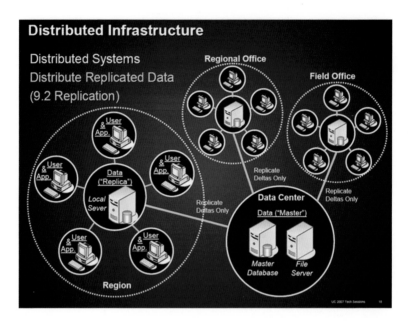

Figure 4.107 Distributed GIS structure of the future. (From Esri, Redlands, California.)

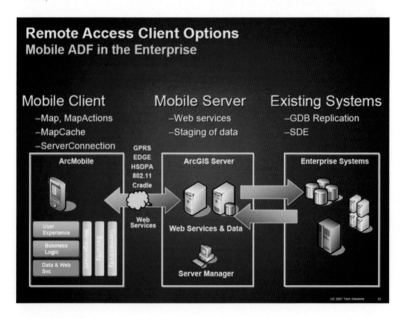

Figure 4.108 Remote access options. (From Esri, Redlands, California.)

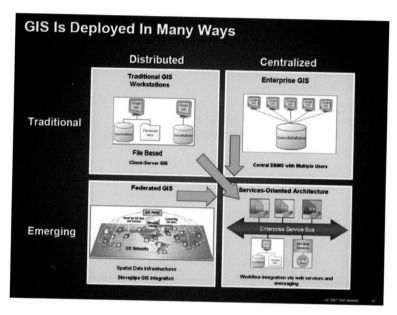

Figure 4.109   Tendency toward service-oriented infrastructure. (From Esri, Redlands, California.)

Figure 4.110   ArcGIS Online. (From Esri, Redlands, California.)

The data can be superimposed with demographic maps, land use maps, and a world terrain map.

Other tools are for 3D city models: From 2D building footprints and the number of floors, 3D city models may be created with standard texture for different building zones.

For critical areas, multispectral Digital Globe WorldView 2 imagery (60 cm GSD) and multitemporal RapidEye data (6 M GSD) will become accessible through a Premium Services Platform, in which a search for the most recent imagery is possible.

Esri's GIS competitors are Autodesk, Bentley, Intergraph, Map Info, Oracle Spatial, SICAD, and Smallworld. Autodesk has targeted the design applications in mechanical engineering, architecture, and engineering constructions, for which, due to their 3D complexity, shapefile use was not relevant. Autodesk is a leader in CAD systems.

Bentley, much like Autodesk, was preoccupied with CAD system uses. For mapping it was the originator of Microstation, which served as a CAD base for Intergraph products until it was replaced by Intergraph's Imagineer for Intergraph's Geomedia product. Since then Bentley developed its own system, called Geographics.

Intergraph, founded in Huntsville, Alabama, in 1969, was originally a hardware-oriented company, which, in the 1980s was able to provide GIS capabilities to the military and plant design market at a time when pure software-based solutions, such as ArcInfo, were not able to meet the computer performance limitations for large databases. Intergraph's early IGDS-DMRS solution of the 1980s based on Bentley's Microstation was redeveloped in the 1990s into the MGE environment. Spatial analysis capabilities with a partial topological structure were added with MGA. The integration of raster data was made possible with IRAS. For the treatment of large volumes of data, MGDM was developed, which could split large area coverages into domains consisting of selected themes. The product TIGRIS developed for large military databases with topological structure found its way into the civilian world under the name of DYNAMO. Finally, the graphic CAD engine of Geomedia was replaced for the Geomedia products of Intergraph as a competition to ArcInfo. Intergraph has now been purchased by the global company Hexagon.

Map Info was founded in 1986. It was purchased by Pitney Bowes. It is a competing product to ArcGIS.

Oracle Spatial and Graph (formerly Oracle Spatial) is a special option of the Oracle database for GIS applications. It contains a topology data model and supports TINs, point clouds, and lidar data. It has special analysis capabilities and is Web compliant. Its advantage is spatial indexing of the database, which offers fast access to the data contained in huge geodatabases.

SICAD/Open was being developed by the German company SIEMENS in 1979. After early applications in facilities management and in the European cadastre, the follow-on company AED-SICAD still maintains SICAD/Open systems but has switched to customization tasks for Esri ArcGIS.

Smallworld, established in 1989 in the United Kingdom, bought by General Electric and later by Pitney-Bowes, has the concept of real-world objects, and like ArcGIS uses 2D topology. Similarly, it has provided access to large geodatabases such as SAP and Oracle Spatial.

GIS APPLICATIONS

GIS applications are scale dependent.

Global Applications

The first topographic base data for global applications was the digital chart of the world (DCW) compiled by the U.S. National Imaging and Mapping Authority (NIMA) and distributed at low cost through the U.S. Geological Survey (USGS) at the scale of 1:1000000. It has been integrated from available national data sets. DCW data can be downloaded as Vmap O data in ESRI shp format from GIS Lab at www.gislab.ru. Digital elevation models, also compiled by NIMA, are downloadable at 1×1 km spacings through NOAA: www.ngdc.noaa.gov/mgg/topo/globe.html.

Based on these references, a variety of satellite image data are rectified, monitoring the atmosphere, the land, and the oceans. Typical global data sets are the NDVI for the monitoring of vegetation (see http://earthobservatory.nasa.gov).

On a continental level, Esri, in cooperation with the World Resources Institute (WRI), has provided the African Data Sampler, in which various small-scale thematic coverages can be compared. Satellite images and thematic vector information can be merged for information at the global level in the form of a digital atlas (see www.earth-info.org).

To permit global access to geographic data for the actions of international organizations and continental activities, a global spatial data infrastructure (GSDI) is required. For its implementation, a GSDI international committee structure has been created. According to an "SDI Cookbook" available through www.gsdi.org, the exchange of data through world base clearinghouse nodes is in preparation. Initiatives such as the "Global Map" (1:1000000) and "Digital Earth," propagating the technology, have progressed. They are supported mainly by the United States, Canada, Australia, and Japan, with European countries, China, and others participating. The link to the former

United Nations Cartographic Conferences and the present United Nations Global Geospatial Information Management (UN-GGIM) has permitted the establishment of regional committees, such as the Asia-Pacific SDI committee (GGIM-AP) and a Latin American SDI committee (GGIM-Americas).

National Applications and Spatial Data Infrastructure

The national supply of base data depends on the activities of the national mapping agencies at medium scales (1:25000 to 1:50000), which permit a more detailed GIS analysis. Since these agencies only provide the topographic base data, which must be supplemented and integrated with thematic data sets from other organizations, the lead agency for the exchange of data must strive for the establishment of a national spatial data infrastructure (NSDI) framework. In the United States, a Federal Geographic Data Committee (FGDC) was formed in 1990, and an NSDI was introduced in 1994 by a Presidential Executive Order (no. 12906).

An NSDI not only contains the geodata in various forms (vector data of digital maps, raster data as digital orthophotos and as digital elevation models) but also the metadata describing the data sets. The clearinghouse provides access to them and constitutes a catalog of the data. Parts of the NSDI are also the agreed framework agreements for the provision and updating of the different data sets according to agreed upon standards.

In most European countries and in Canada, such national spatial infrastructures have been established by the national mapping agencies (e.g., Ordnance Survey in the United Kingdom, IGN France, Geomatics Canada) in liaison with user ministries. In countries where the large-scale mapping responsibilities have been assigned to the states of the country, the state survey administrations have created an NSDI by coordination committees (e.g., in Germany by the ADV). In Germany, the national topographic information system ATKIS was completed for the entire country. Its geometry is derived from the digitization of the 1:5000 base maps (DGK) with an accuracy of ±3 m on the ground. So far, the system contains 90 object classes. It is possible to derive map data at a scale of 1:10000 from the data sets covering the country.

Figure 4.111 shows the GIS-derived topographic map, 1:25000, extracted from ATKIS data. It is now integrated with the land information system based on the cadastre ALKIS and the geodetic reference system data AFIS.

These national NSDI efforts have been instrumental in creating the European SDI "INSPIRE," which has been promoted by the European Environmental Agency in Copenhagen and can be considered as a European standard. The base data provided and maintained by the national (or state) mapping agencies

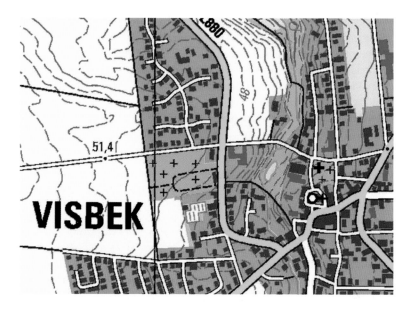

Figure 4.111   The German topographic map 1:25000 derived from ATKIS. (Map courtesy of LGN, Hannover, Germany.)

are being augmented by national, state, or local authorities with their relevant thematic data for their uses for

- Regional planning
- Environmental management
- Statistical purposes
- Soil surveys

Since the vector base data are more costly and less easy to manipulate than raster data, the mapping agencies have also supplied their raster scanned maps at scales of 1:50000, 1:25000, 1:10000, and 1:5000. These data sets are easily available to local authorities and private companies, and they can create a value-added digital map for their specific purposes. They retain the responsibility for maintenance and updates of their data sets according to the agreed exchange standards.

A great number of value-added applications are developable on a project basis such as:

- Hydrological modeling (using vector data, raster images, and DEMs)
- Car navigation (with the need to provide monthly updates for traffic routes)
- Business analysis (based on income of residents)

- Accident and crime location (of interest to police and traffic planners)
- Real estate valuation and real estate market
- Emergency planning (firefighting access, flooding)
- Tourist information systems (location of stores, restaurants, public places of interest, mailboxes, pharmacies, petrol stations, public phone booths, hospitals)
- Health studies (occurrence of diseases in certain locations)

Local Applications

Local decisions most often require large-scale information at map scales ranging from 1:100 for construction to 1:500 for utility location to 1:1000 for urban property cadastres to 1:2000 for rural property cadastres.

Cadastral Applications

An application of survey and geoinformation technologies, in which the integration of survey and date administration technologies is of importance, is the establishment and the maintenance of the real estate property cadastre.

A real estate property cadastre consists of two parts:

1. A cadastral map, in which the geometrically defined and geocoded land objects (land parcels, buildings) are documented. In modern times, this geometric definition may also be in the form of coordinates for each boundary point. The land object is furthermore identified by an object number (parcel numbers, building numbers).
2. A land register, in which all alphanumeric data are contained for the attributes of each land object (owner's name, his ID, land use, mortgage, type of land right). For query purposes this register is best organized as a relational database.

In historic times this was realized as a land parcel map and a land register containing handwritten entries or containing documents such as deeds or titles.

A first systematic introduction of a cadastral system was done in the Austro-Hungarian Monarchy in the so-called Josephine cadastre around 1780.

Prerunners of a cadastral survey without maps were the Doomsday Book of 1086 in England and the establishment of the Landmäteriet in Sweden in 1628 for the purpose of taxation.

In England and its early American dependencies the "metes and bounds system" to describe land parcels without geocoding and topological relations became practice. In the U.S. Western states, the old settlements land became distributed to settlers according to the Public Land Survey System, but a cadastral system like the one introduced by Napoleon in France in 1807, which spread cadastral systems in Europe, was never introduced.

In Europe, the cadastral systems have gone through a transition from the "tax cadastre," where only the area of the parcel was the basis of taxation to the "ownership protection cadastre" of the early 20th century with an overemphasis on relative accuracy to the "multipurpose cadastre" started in the 1930s, by which it was possible to serve the planning needs of the community within reasonable accuracy limits.

In the last decades the International Federation of Surveyors FIG, and specifically its commission 7 on the cadastre, has targeted the need for the establishment and the maintenance of cadastral systems by modern technology. In 1994, the vision of "cadastre 2014" was developed. It started with the fact that only about 50 countries of the 200 on the globe have established efficient land registration systems, and 50 are still in the process to establish one, while another 100 cannot take advantage of its existence.

The Peruvian economist Hernando de Soto has expressed that "insecure land right deprive a country of the capacity of land to serve as collateral for mortgages. There land becomes dead capital, and this has serious handicaps for economic development." For this reason the World Bank has supported the establishment of cadastral land registration systems on the globe.

Modern methods to introduce cadastral land registration systems have been used in the transformation counties of the Balkan and of the former Soviet Union, where land property ownership had become abolished over a period of 50 to 70 years. There the privatization of agricultural land has proceeded in the following steps:

1. Former state agricultural land was distributed to citizens by means of certificates.
2. Local survey companies were contracted to survey and to optimize the granted fields on the ground. Contrary to survey practices 100 or more years ago this could already be done using GNSS.
3. The government instituted orthophoto imaging programs for the entire country to serve as quality control for the locally made surveys, for which 10% to 20% of mistakes became visible.

In areas with smaller and irregular parcellation, methodologies practiced in Georgia in the Caucasus region and in Cambodia have proved successful. A local survey team equipped with digital orthoimages on tablet PCs initiated a process of local adjudication between neighbors. They photographed the deeds of the owners by their digital cameras, then they surveyed the agreed boundary points by GNSS phase receivers in RTK mode. The survey results went by mobile phone to the head office, where the map and attribute database was created.

Some African countries, where orthophotography would be too costly an adaptation of this method using Google Earth images has been suggested.

Where available, the cadastre served as a register of property for the purposes of property tax collection. It consisted of a large-scale map indicating the geometric location of land parcels with their parcel numbers. In the accompanying register, the property owners or users were recorded with a cross-reference to the parcel numbers of each district, village, and block. The capital market, which permitted the owners of land to obtain loans on the basis of mortgages, made the cadastre a valuable asset to record these, together with land use rights and encumbrances in the nongraphic part of the cadastre (the "book"). The graphic part of the cadastre (the "map") permitted the administration of the neighborhood relations of parcels, giving the owner the guarantee that the land parcel existed within the indicated dimensions without overlaps. Even though property boundaries were also monumented, the cadastral map proved useful in retracement surveys of these boundaries, in case monuments were lost.

In the early 20th century, these property boundary points were tied into geodetic reference systems. The field survey records permitted the recalculation of the exact coordinates of the boundary points for the reestablishment of lost boundary markers. This permitted the establishment of a numerical cadastre of the parcel geometry in some densely populated regions, as opposed to a purely graphical cadastre, in which the boundaries were only recorded on the large-scale map.

The historical development of cadastral systems was different in the various regions of the world. In Central Europe, a numerical or at least a graphical cadastre was established, whereas the English-speaking world favored land title systems in which the geometrical references were contained in deeds or title documents.

With the advent of GIS, both types of property description and registration could be converted into a vector-based digital form. The cadastral GIS now contains graphical data in georeferenced form, describing the location and the dimensions of parcels and nongraphic attributes containing references to owners, land use, and tax value. Emphasis is on a unique identifier, the parcel number, which is quickly locatable by the coordinate references. A cadastral GIS can now serve the purposes of

- A tax cadastre
- An ownership protection cadastre with the function of a land titles system
- A multipurpose cadastre for the purpose of planning the local environment

In Germany, the digital cadastral map ALK has been established on a 1:1000 or 1:2000 scale map base, uniquely identifying the parcel. The boundary points are georeferenced in point coordinate registers. A separate database exists for parcel numbers with the attributes (ownership, etc.) in the form of ALB. It is possible to obtain all relevant parcel records at the cadastral offices within

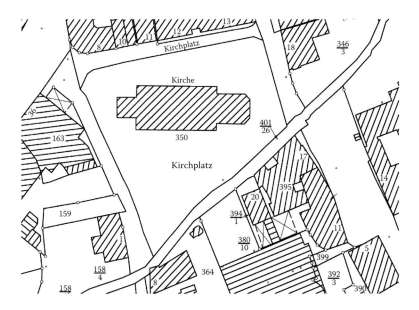

Figure 4.112   Cadastral map 1:1000. (Map courtesy of LGN, Hannover, Germany.)

minutes. By the year 2007, the ALK and the ALB were merged into a new cadastral GIS, with the name of ALKIS, for the entire country.

Figure 4.112 shows a part of the ALK map, 1:1000, depicting land parcels and boundaries, boundary points, buildings, land use, annotated land use, and a limited number of topographic objects. Figure 4.113 shows an ALKIS ownership certificate. Figures 4.114, 4.115, and 4.116 show value added applications of the cadastral GIS data for urban mapping.

The procedure for generating value-added information is also shown in Figures 4.117, 4.118, and 4.119 in an example to derive access information for fire engines.

Recognizing the value of such a system for the economic stability of a country, the World Bank has supported many land management projects worldwide, which all contain land information systems on the basis of some form of digital cadastre. Typical examples are the Land Titling projects in Thailand with Australian consultancy; in Peru with Canadian consultancy; and in the reform countries of Eastern Europe with American, Scandinavian, and German consultancies. These concentrate on low-cost approaches. If property or land use records do not exist, the location and the dimensions of parcels are established by photo adjudication, in which neighboring landowners or land users agree on boundary locations identified in photographs or in orthophotos. Figure 4.120

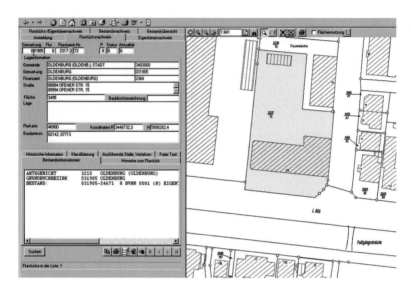

Figure 4.113 Ownership certificate. (Map Hannoversch-Münden (ALK), courtesy of VKV Niedersachsen, Dezernat 207, Bez.-Reg. Weser-Ems, Oldenburg, Germany.)

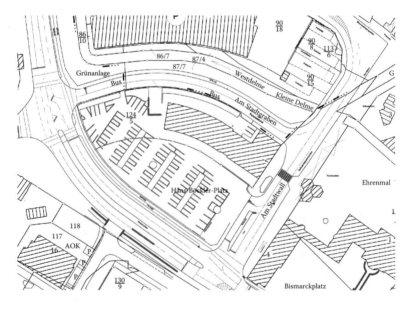

Figure 4.114 Cadastral map with road topography. (Map Delmenhorst, courtesy of VKV Niedersachsen, Dezernat 207, Bez.-Reg. Weser-Ems, Oldenburg, Germany.)

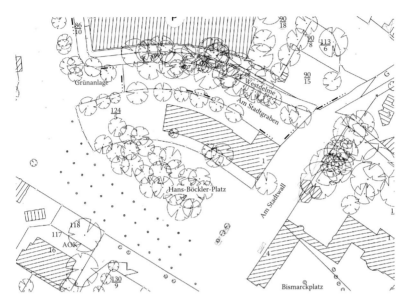

Figure 4.115 Cadastral map with urban vegetation. (Map Delmenhorst, courtesy of VKV Niedersachsen, Dezernat 207, Bez.-Reg. Weser-Ems, Oldenburg, Germany.)

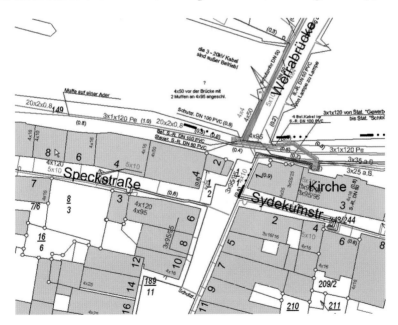

Figure 4.116 Cadastral map with utilities. (Map Delmenhorst, courtesy of VKV Niedersachsen, Dezernat 207, Bez.-Reg. Weser-Ems, Oldenburg, Germany.)

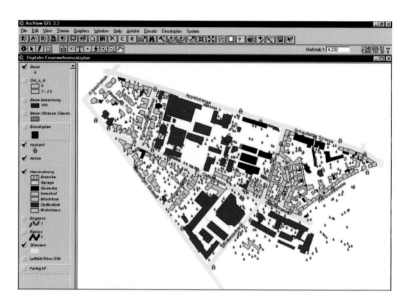

Figure 4.117   Cadastral GIS with minimal topographic information. (Map courtesy of Institute for Photogrammetry and GeoInformation, University of Hannover, Germany.)

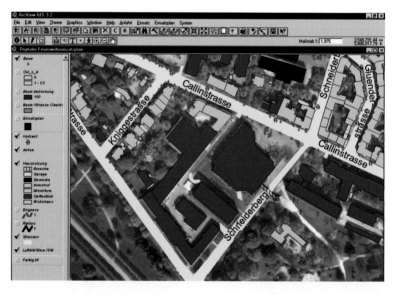

Figure 4.118   Digital orthophoto for deriving crown diameters of trees. (Map courtesy of the Institute for Photogrammetry and GeoInformation, University of Hannover, Germany.)

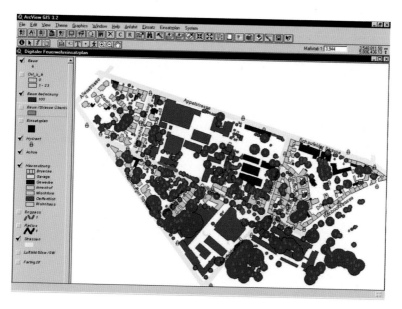

Figure 4.119   GIS with crown diameters of trees. (Map courtesy of the Institute for Photogrammetry and GeoInformation, University of Hannover, Germany.)

shows the superposition of digital orthophoto with the relatively old cadastral map in Croatia with the need for the redetermination of boundaries.

Contrary to topographic GIS databases, which have a periodic updating need at intervals of a few years, cadastral GIS databases must be updated in near-real time. This imposes strict administrative procedures and efforts on behalf of governments assuring a secure land management for public and private uses.

Facility Management

Another special GIS application area is in the inventory, management, and planning of the utility infrastructure concerning the distribution of electricity, telephone, and computer line networks; the supply of fresh and irrigation water, and of natural gas; and the management of sewerage and rain water drainage systems.

A GIS permits the location of the manholes and the specific devices (distributor boxes, transformers, valves, etc.) geometrically. These are interconnected by cables or pipes in a linear network. Each utility network can be contained in separate graphic layers, which may be superimposed in a GIS, describing their relative location. Figure 4.121 shows an urban utility network.

Figure 4.120 Superposition of old cadastre and digital orthophoto in Croatia. (Map courtesy of State Geodetic Administration (DGU), Republic of Croatia, Zagreb.)

Often, the utility administrators or companies prefer a schematic display of individual cables indicating their switching capabilities and their capacities. Attributes may describe depth of cables and pipes or their relative slope. Cable and pipelines may be linked into object-oriented networks, permitting the analysis of flows. Even the consumption of power, gas, or water may be queried in these networks up to the point of billing.

Figure 4.121 Urban utility network. (Map courtesy of the Ministry of Municipal Affairs and Agriculture, Doha, Qatar, C.E. Department, Drainage GIS Section, and is used herein with permission by Esri, Redlands, California, and the Ministry of Municipal Affairs and Agriculture, Doha, Qatar.)

City Models

A relatively new requirement for the use of three-dimensional information from a GIS has developed from the mobile telephone industry. To locate the optical distribution of cellular phone antennas in urban areas, the mobile telephone industry needs three-dimensional city models to circumvent transmission obstructions caused by buildings. A standard base data set obtained by analogue, analytical, or digital photogrammetry contains the 2D dimensions of the building footprints at the ground level of a topographic DEM. Additional manual photogrammetric measurements of the rooftops, which may also be obtained in an automated way by image matching techniques, permit the generation of 3D city models at various stages:

- The company Phoenics of Hannover has generated about 100 city models of major German cities in a block form. Building polygons have been extracted from Atkis data. The building heights have been added via image correlation. This permitted the creation of the 3D building model as an input to the mobile phone antenna location programs.

- These models can be improved by the additional measurement of several roof types to permit schematic oblique views of the city.
- It is possible to integrate texture images taken by terrestrial or airborne handheld small digital cameras by a rectification process to the façades of the buildings with the Photomodeller program. This "beautification process" is useful for city centers or the most prominent buildings (public buildings, churches, railway stations).
- It is then possible to develop flythroughs through the cities for the creation of animations with the Erdas Virtual GIS program.

Figures 4.122 to 4.126 show the sequences of operation in obtaining a 3D city model:

1. Digital surface model (DSM) obtained by image correlation from photogrammetry
2. Digitization of buildings from a vector GIS
3. DEM from image correlation subtracting building area signals
4. Creation of building blocks after vector digitization of building outlines using height differences between DSM and DEM for the house areas; superposition of the digital orthophoto onto the ground
5. Generation of roof structure from DSM height level differences for the buildings; pasting of texture information to building façades from handheld digital camera images

The company Phoenics-Vectuel of Hannover has used 3D city models, enhanced with rendering information for planned construction and vegetation,

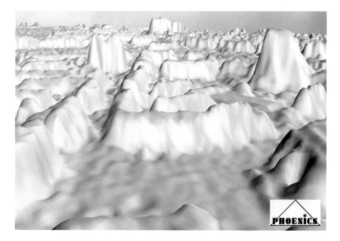

Figure 4.122 DSM via image correlation. (Image courtesy of Phoenics GmbH, Hannover, Germany.)

Figure 4.123   Extraction of buildings from a GIS. (Image courtesy of Phoenics GmbH, Hannover, Germany.)

for the display and assessment of the urban landscape development alternatives (Figures 4.127 to 4.130).

The significance of a cadastral system is to cover the land areas of a community, a country, or the entire globe with a uniquely geometrically defined and geocoded fabric of land parcels that do not overlap and that cover the entire

Figure 4.124   DEM via image correlation after subtraction of buildings. (Image courtesy of Phoenics GmbH, Hannover, Germany.)

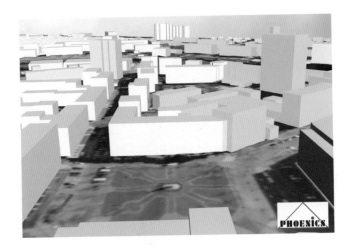

Figure 4.125   Creation of buildings from different heights DEM–DSM. (Image courtesy of Phoenics GmbH, Hannover, Germany.)

cadastral area without gaps. This can be achieved if all land parcels of the cadastral area are defined by their boundary points defined by their geodetic coordinates in a predefined geodetic reference system (Figure 4.131). In general, the reference system is a reference ellipsoid tied to a terrestrial reference frame, tied to an astronomic reference frame at a certain epoch.

To permit a neighborhood analysis of parcels, the boundary topology with respect to the neighboring parcels also has to be recorded. The area features of the land, defined as parcels or as buildings or as other units (environmental protection zones), are to be uniquely defined by a parcel number (Figure 4.132), building number, or special area number. In Britain, the identifier for such a defined area is called a TOID (Figure 4.133). This parcel number, building number, or area number is the link to alphanumeric information describing the attributes of the parcel, building, or area, for example, the name of the owner or user, his identification, the right he has with respect to use or ownership, the use of the parcel, and the date of the entry of the record. All attributes are best administered through a relational database, the core of the juridical land registration system.

When both the attribute content and the geometrical parcel fabric are administered in a GIS, then it becomes possible to utilize the cadastre and the land information system for purposes of land management (see Figure 4.134). This type of geometric description in the database by referenced coordinates makes it unnecessary to record field measurements by metes and bounds in the survey plans, as this is the practice in the United States (see Figure 4.135).

Figure 4.126   Addition of building façade texture, orthophoto texture for the ground and derivation of roof types from DSM height differences for buildings. (Image courtesy of Phoenics GmbH, Hannover, Germany.)

Figure 4.127 Virtual reality application (present). (From Phoenics-Vectuel, Hannover, Germany.)

Figure 4.128   Virtual reality application (future). (From Phoenics-Vectuel, Hannover, Germany.)

Instead, both the land registration documents (see Figure 4.136) and the surveyed object topography (Figure 4.137) may be used as an output of choice.

A further possibility (Figure 4.138) practiced in land management is possible when merging high-resolution satellite images or aerial orthoimages from aerial surveys showing the constructed buildings superimposed with the

Figure 4.129   Virtual reality application (option 1). (From Phoenics-Vectuel, Hannover, Germany.)

Figure 4.130 Virtual reality application (option 2). (From Phoenics-Vectuel, Hannover, Germany.)

Topologically Consistent Data for Areas Surveyed with Boundaries

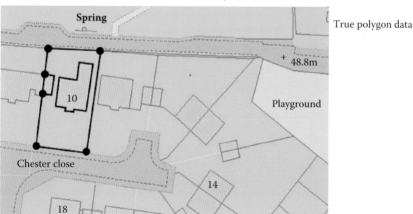

Figure 4.131 Recommended cadastral surveys with boundaries. (From Ordnance Survey, Southampton, United Kingdom.)

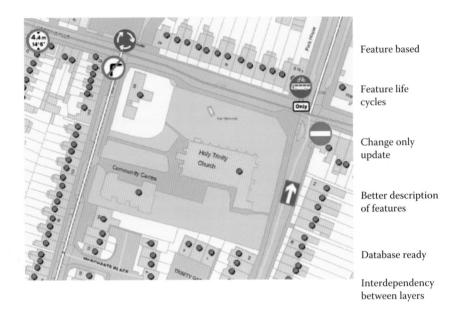

Feature based

Feature life cycles

Change only update

Better description of features

Database ready

Interdependency between layers

Figure 4.132 Inclusion of feature points. (From Ordnance Survey, Southampton, United Kingdom.)

Area Features are Identified with a Unique Number, the TOID

Spring

Unique feature IDs (TOIDS) and version numbers

+ 48.8m

TOID: 0001100029000372
Version: 2

10

Playground

TOID: 0001100029009055
Version: 2

Chester close

14

18

Figure 4.133 Identification of features by unique number. (From Ordnance Survey, Southampton, United Kingdom.)

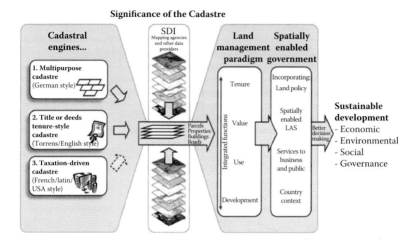

Figure 4.134 Use of cadastral information for land management. (From Esri, Redlands, California.)

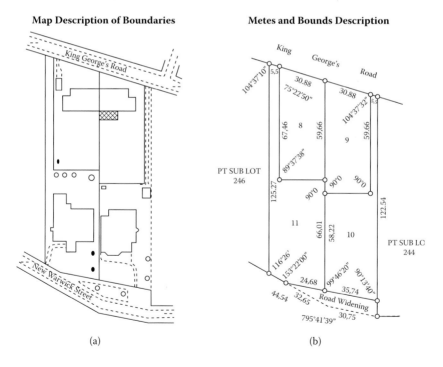

Figure 4.135 Metes and bounds description versus coordinate description of boundaries from own old papers.

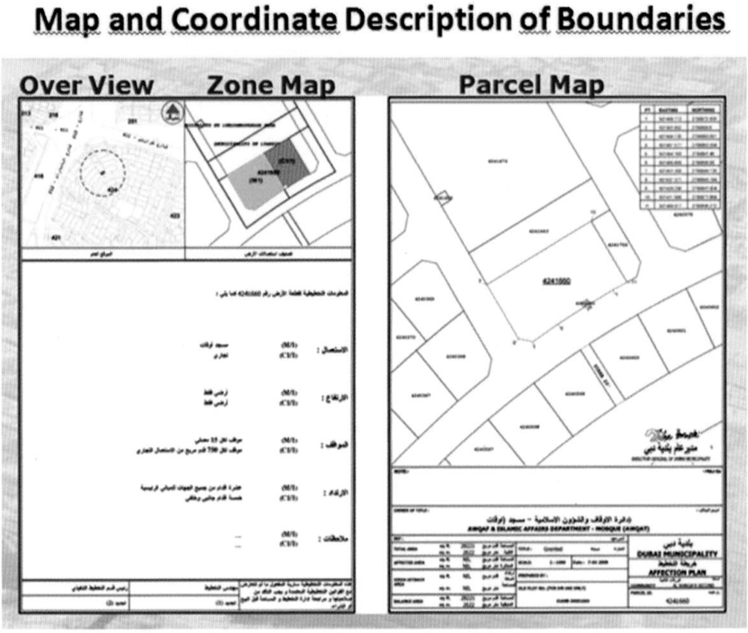

Figure 4.136   Land registration document. (From Ordnance Survey, Southampton, United Kingdom.)

information in settlement plans to show that only part of the planning scheme had been completed at the time of the imagery.

Spatial Analysis Applications

There is a wide variety of spatial analysis applications, which are demonstrated by the following examples.

Figure 4.139 shows a demographic analysis for statistical data obtained for the counties of the United States.

Figure 4.140 shows a municipal application for the different governmental, residential, and commercial housing in the city of Bangkok.

Figure 4.141 shows ownership of parcels in individually owned, business owned, and institutionally owned land in a U.S. city.

Figure 4.142 shows the occurrence of a beetle infestation in New York City by geographic location.

Figure 4.143 shows a road network with the analyzed travel distance to the nearest fire station in a U.S. city (black = less than 1.1 miles, red = 4.3 miles).

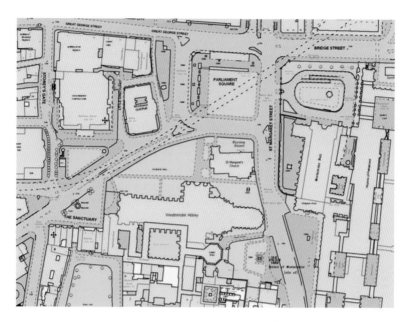

Figure 4.137   Surveyed object topography turned into a map. (From Ordnance Survey, Southampton, United Kingdom.)

Figure 4.138   Merging up-to-date imagery with planning information to monitor construction progress screenshot.

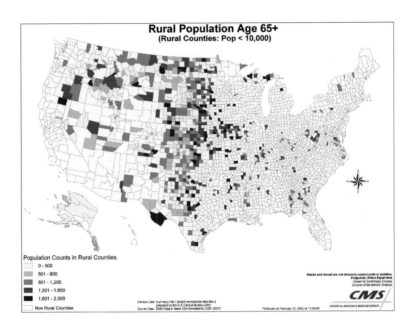

Figure 4.139   Rural population with age over 65 in counties in the United States. (Map courtesy of Applied Geographics Inc., Boston, Massachusetts, and is used herein with permission by Esri, Redlands, California, and Applied Geographics.)

Figure 4.144 shows the frequency of burglaries by blocks in a U.S. city.

Figure 4.145 shows the extent of industrial air pollution in a city (light blue = very low, dark blue = heavy).

Emergency GIS over Crisis Areas

The following describes the compilation of a crisis area GIS of the Kosovo region of Serbia. Figure 4.146 shows the layers of the crisis management GIS:

- Satellite image database
- Topographic NATO map
- Additional GPS-related information
- Digital elevation model
- European CORINE land cover map

Figure 4.147 shows the ERS 1/2 radar interferometry generated DEM of the area. Figure 4.148 shows the satellite image database draped over the DEM.

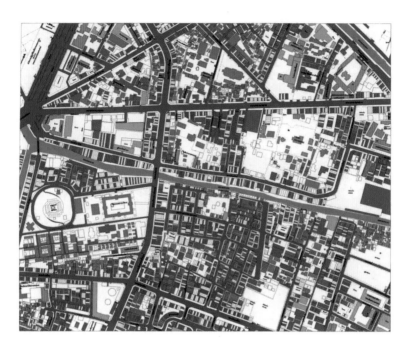

Figure 4.140 Governmental, residential, and commercial holdings in the city of Bangkok. (Map courtesy of Metropolitan Electric Authority, Thailand, and is used herein with permission by Esri, Redlands, California, and Metropolitan Electric Authority for educational purposes.)

The use of remote sensing is particularly useful in developing countries. As a demonstration, two examples from Kenya are depicted. Figure 4.149 shows a section of the governmental map 1:50000 of Nairobi. It depicts the slum area of Kibera, where close to 1 million people live, as an empty area, since the occupation of the land is illegal. Figure 4.150 is the corresponding Google Earth image of Kibera, taken by a high-resolution satellite. Figure 4.151 shows the supply of drinking water in the area, and Figure 4.152 is an image of the sewerage system. Figure 4.153 shows the attempt to assess the sanitary conditions in Kibera by a GIS.

Another application of remote sensing data is the use of medium-resolution satellite imagery (Landsat) to study the interference of growing settlements with local agricultural activity (Figure 4.154) with wildlife tracks between the game areas of Ambo Seli and the Nairobi National Park (Figure 4.155). The local observation is that sheep and zebras cross their paths (Figure 4.156).

Figure 4.141   Ownership type map of a U.S. city. (Map courtesy of Middle Rio Grande Conservancy District, Albuquerque, New Mexico, and is used herein with permission by Esri, Redlands, California, and the Middle Rio Grande Conservancy District.)

Figure 4.142   Beetle infestation in New York City. (Map courtesy of the City of New York Parks and Recreation, New York, New York, and is used herein with permission by Esri, Redlands, California, and the City of New York.)

General Spatial Information Systems

An example of tourist information systems is shown for a Tourist GIS made commercially available by the company Geospace. It displays, on the basis of a digital photomap,

- Street names
- House numbers
- Telephone booths
- Restaurants
- Hotels
- Tourist sites
- Pharmacies
- Hospitals and other relevant information (Figure 4.157)

Figure 4.143 Distances to the nearest fire station. (Map courtesy of Montgomery County, Rockville, Maryland, and is used herein with permission by Esri, Redlands, California, and Montgomery County.)

Figure 4.144 Frequency of burglaries by city blocks in a U.S. city. (Graphic image courtesy of Esri-ArcView GIS, and is used herein with permission by Esri, Redlands, California.)

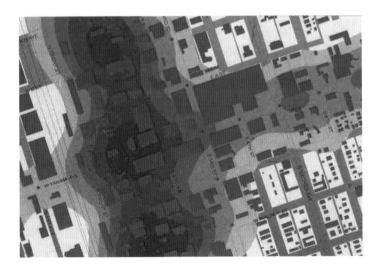

Figure 4.145 Industrial air pollution in a city. (Map courtesy of the City of Yakima, Washington, and is used herein with permission by Esri, Redlands, California, and the City of Yakima.)

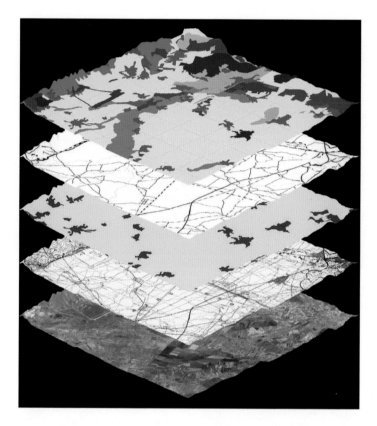

Figure 4.146   The layers of a crisis management GIS over Kosovo. (Landsat, source: P. Reinartz et al., in PGM 2000/3, processed by DLR, courtesy of DLR, Oberpfaffenhofen, Germany.)

Such efforts have been brilliantly taken over by systems such as Google Earth and Bing Maps. Google Earth is a software system based on a virtual globe to visualize images and superimposed map features of the earth's surface. In its simplest form it is available free of charge to the Internet user. Figure 4.158 shows the initial view of the Globe for search of the desired detailed viewing area by place name or coordinates. Figure 4.159 shows a portion of the Google Earth content for the building of the Institute for

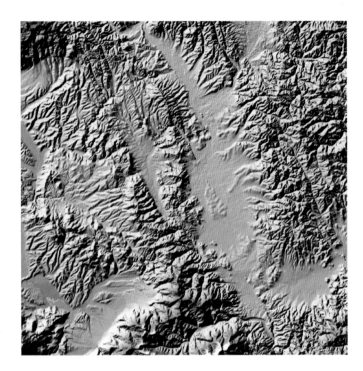

Figure 4.147 The DEM over Kosovo, generated from radar interferometry. (DLR and JRC, processed by DLR, courtesy of DLR, Oberpfaffenhofen, Germany.)

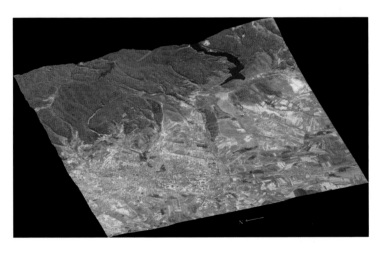

Figure 4.148 Satellite imagery draped over the DEM. (DLR and JRC, processed by DLR, courtesy of DLR, Oberpfaffenhofen, Germany.)

NAIROBI

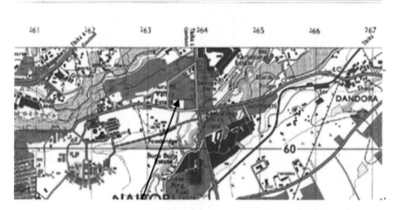

Figure 4.149   Section of topographic map 1:50000 of Nairobi over Kibera. (From the Department of Geospatial and Space Technology, University of Nairobi.)

Photogrammetry and Geoinformation of Leibniz University Hannover in Germany. The panning and zoom capabilities of the viewer permit viewing of the data at the desired resolution (Figure 4.160 shows the zoomed-in image of Figure 4.159).

Google Maps is a simplified map content of the areas selected for view in Google Earth (Figure 4.161). Imagery and maps can be superimposed to show map text on the images, for example, street names (Figure 4.162). Google Earth 3D is an added option to visualize the image content in three dimensions. This has been introduced for urban centers with high-rise buildings (see Figure 4.163 for the city hall area of the city of Hamburg, Germany). Google Street View has been created for all areas of the United States and of Western Europe to image terrestrial street views along the urban road networks contained in Google Earth and Google Maps (see Figure 4.164 for the street view of the building of the Geodetic Faculty Hannover).

Figure 4.150   Google Earth image of Kibera. (From the Department of Geospatial and Space Technology, University of Nairobi.)

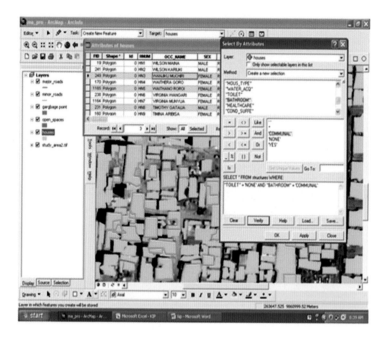

Figure 4.151   ArcGIS application for Kibera. (From the Department of Geospatial and Space Technology, University of Nairobi.)

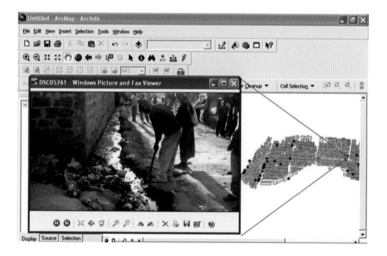

Figure 4.152   Drinking water in Kibera. (From the Department of Geospatial and Space Technology, University of Nairobi.)

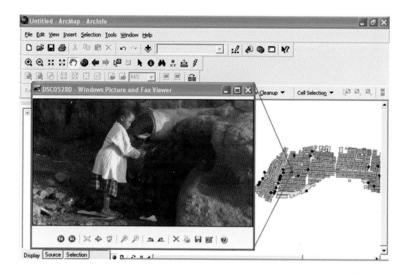

Figure 4.153   Sewerage in Kibera. (From the Department of Geospatial and Space Technology, University of Nairobi.)

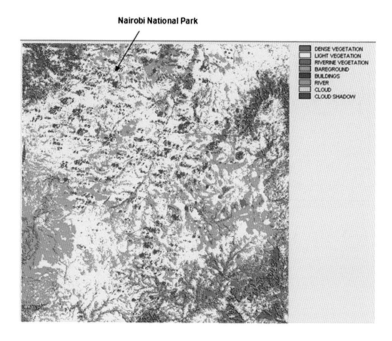

Figure 4.154   Landsat classification of wildlife conflict area south of Nairobi Park. (From the Department of Geospatial and Space Technology, University of Nairobi.)

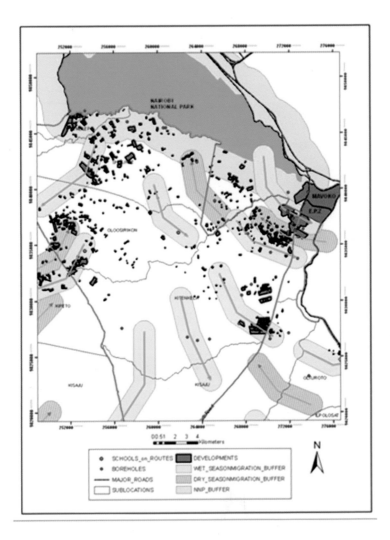

Figure 4.155 Map of game movements. (From the Department of Geospatial and Space Technology, University of Nairobi.)

Figure 4.156 Conflicts between animals. (From the Department of Geospatial and Space Technology, University of Nairobi.)

Figure 4.157 Example of a tourist GIS for the city of Bonn, Germany. (Image courtesy of Geospace GmbH, Köln.)

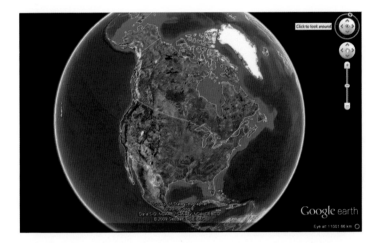

Figure 4.158   The Google Earth Globe Viewer.

Figure 4.159   Google Earth view of the University of Hannover, Germany.

Figure 4.160   Google Earth zoom view of the University of Hannover, Germany.

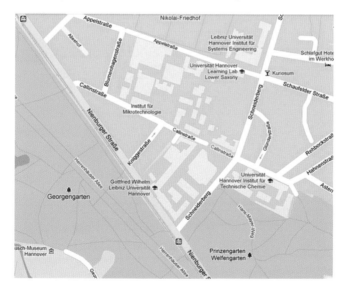

Figure 4.161   Google Maps view of the University of Hannover, Germany.

Figure 4.162   Google Maps view of the University of Hannover, Germany: superimposed imagery.

Figure 4.163   Google Earth street view of the University of Hannover, Germany.

Figure 4.164   Google Earth 3D view of Hamburg City Hall, Germany.

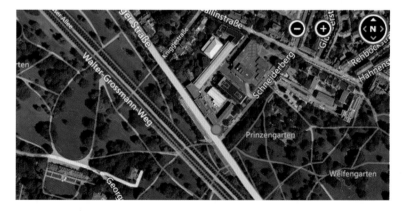

Figure 4.165   Bing Maps view of the University of Hannover, Germany: superimposed imagery.

Figure 4.166   Bing Maps view of the University of Hannover, Germany.

Microsoft has introduced Bing Maps as a competitive product to Google Earth (Figure 4.165) and to Google Maps (Figure 4.166). In China, similar developments have recently taken place. The LIESMARS laboratory of Wuhan University has developed GeoGlobe as a Chinese Google Earth, and so has the company Map World (Tianditu).

Chapter 5

Positioning Systems

The capability to geocode GIS information in vector and raster form today depends on modern geopositioning systems.

THE GLOBAL POSITIONING SYSTEM (GPS)

The NAVSTAR–GPS (Navigation System with Time and Ranging–Global Positioning System) has been under development by the U.S. military since 1973. From 1993, the system has consisted of 21 active satellites and 3 spares. The satellites orbit the earth at an altitude of 20200 km. Their orbits are arranged in such a manner that at least four of them are in direct line of sight at any point of the earth's surface, 24 hours a day, unless the line of sight is obstructed by buildings, vegetation, or steep topography. The use of GPS has been developed for real-time navigation and for geodetic positioning. Figure 5.1 illustrates the global GPS satellite configuration.

GPS Signals

The satellites transmit signals in two carrier frequencies: L1 at 1575.42 MHz and L2 at 1227.60 MHz. Onto these carrier waves, specific codes are modulated that permit the measurement of distances from a particular satellite to a receiving antenna on a GPS receiver. Figure 5.2 shows the modulation of the carrier wave by codes.

For three known satellite positions on a common reference frame, the receiver coordinates can be computed on that reference frame by the three distances measured. A fourth satellite, however, needs to be observed due to the lack of synchronization between satellite clocks and the receiver clock. The ephemeris data of all 34 satellites are continuously determined by the operational network of control stations at Colorado Springs, Colorado; Ascension Island; Diego Garcia; Kwajalein Atoll; and Hawaii.

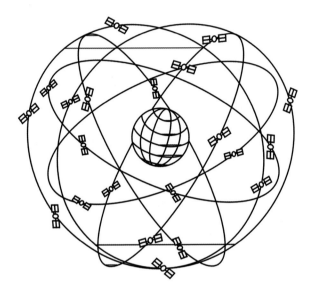

Figure 5.1   Global GPS satellite configuration.

Figure 5.3 shows the determination of the position from the four distance measurements, R_1, R_2, R_3, and R_4.

The measurement of distances between satellite and receiver is made possible by the codes modulated onto the carrier frequencies. There is a *P*-code (precision code) available to military users in encrypted form with a frequency of 10.23 MHz corresponding to a distance of 30 m. Civilian users must utilize the C/A code (clear/acquisition code) with a frequency of 1023 MHz corresponding to a distance of 300 m. The receiver has a copy of the code, which is shifted in

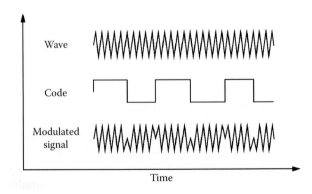

Figure 5.2   Modulation of the carrier wave.

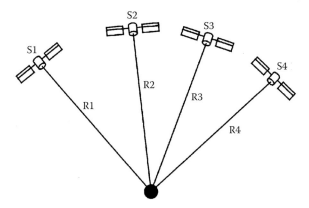

Figure 5.3   Absolute positioning.

time steps and correlated with the observed signal. The *P*-code permits direct distance observations with an accuracy of 0.6 m to 1 m. The C/A code permits the observation of distances with an accuracy of 6 m to 10 m.

Low-cost satellite receivers permit the direct utilization of the code observations. Higher precision and accuracy is achievable with high-cost satellite receivers, which permit the utilization of the phase of the received carrier waves in addition to the C/A code. Depending on the observation time, the constellation of satellites and the atmospheric conditions carrier phase receivers are able to determine receiver positions to an accuracy or at least to a precision of 1 cm to 2–3 mm.

Code and carrier wave signals are receivable every second within a specific data block. Other blocks transmit data on the clock parameters, the broadcast ephemeris of the satellite observed, and, at larger time spacings, the ephemeris data for all available GPS satellites plus ionospheric correction parameters. Up until mid-2000, the C/A code was artificially disturbed up to 100 m by the so-called selective availability limiting the use of low-cost satellite receivers to that accuracy.

Figure 5.4 shows the causes of disturbances of GPS signals. Figure 5.5 assesses the magnitude of these disturbances. Ionospheric delays are the largest error source affecting absolute positioning. For this reason, differential GPS (DGPS) has been introduced as an operational relative observation technology. Even for carrier phase measurements, the DGPS technique has proved useful, since phase measurements are affected by an ambiguity term expressing the unknown number of complete wavelengths in the observed distance. It is deduced from the C/A code. But if it is faulty, a cycle slip by a full wavelength may occur.

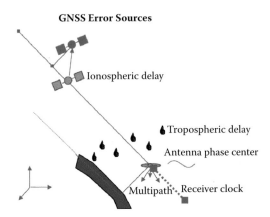

GNSS Error Sources

Figure 5.4   Disturbances of GNSS signals.

Differential GPS

In DGPS, a master GPS receiver is utilized at a geodetic reference station. It receives the same satellite signals observed at a transportable second receiver (rover), which is used for the measurements at new point locations (see Figure 5.6). In this way, relative positioning becomes possible with much higher accuracies. For distances between rover and master receivers of under 10 km, relative coordinate determinations in the range of ±1 cm in position and ±3 cm in ellipsoidal height become possible.

Magnitude of Error Sources

Error source	Absolute influence	Relative influence
Satellite Orbit	2 ... 50 m	0.1 ... 2 ppm
Satellite Clock	2 ... 100 m	0.0 ppm
Ionosphere	0.5 ... > 100 m	1 ... 50 ppm
Troposphere	0.01 ... 0.5 m	0 ... 3 ppm
Multipath Code	m	m
Multipath Phase	mm ... cm	mm ... cm
Antenna	mm ... cm	mm ... cm

High spatial correlation ▮   Local (Calibration) ▮

Figure 5.5   Magnitude of error sources of GNNS signals from own reports for FAO.

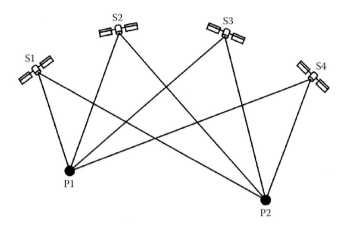

Figure 5.6 Relative positioning.

The disadvantage of using master–rover combinations for DGPS observations lie in the fact that two receivers have to be simultaneously operated. Nevertheless, this has proved feasible for airborne GPS observations for which accuracies of ±10 cm to 15 cm for the exposure station positions could be reached. For terrestrial measurements, there has been the tendency in a number of countries (see Figures 5.7 to 5.10) to establish a number of permanently receiving master stations, which offer DGPS correction through radio transmission or through a mobile GSM network to the rover stations. The German SAPOS network, by the state survey administrations, provides such services with permanent receiver stations located at 50 km intervals.

The error sources for GPS signals stem from the following sources:

- Ionospheric influences, to be reduced by continuously operating reference station (CORS) networks
- Tropospheric influences to be reduced by atmospheric modeling
- Multipath effects at the rover stations, which are station dependent, avoiding observations close to tall buildings or power lines
- Receiver antenna eccentricities, which can be eliminated through calibration

Of these, the ionospheric influences are the most serious. Two German companies—Geo++ of Garbsen, near Hannover and Terrasat of Höhenkirchen, near Munich—have specialized in software systems to model the ionospheric effects from the observations of several permanent reference stations and to transmit the local ionospheric corrections to users, so that they can receive the corrections within less than a minute at the receiver with accuracies in

Figure 5.7   SAPOS coverage for Germany.

the 1 cm range for position and 3 cm for ellipsoidal elevation within the 50 km spaced network of permanent reference stations in the real-time kinematic (RTK) mode of operations. This requires resolution of the ambiguities in static mode before the rover starts moving in phase lock.

On a much wider scale, a permanent reception station is available near Frankfurt (Main), which transmits DGPS corrections via long-wave radio transmission enabling central European users to reach real-time positioning in the 1 m range.

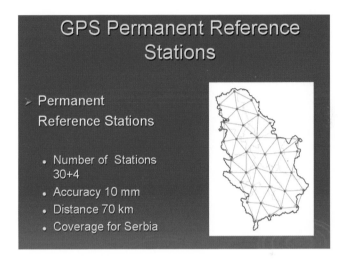

Figure 5.8   CORS coverage for Serbia.

Such international stations are interconnected worldwide to a civilian cooperative international GPS network (CIGNET), which releases global ephemeris information covering an 8-day period through the U.S. Coast Guard GPS information center and has about 20 international tracking sites. The information may be utilized for postprocessing.

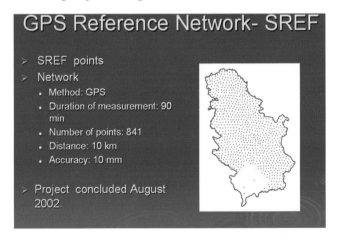

Figure 5.9   Densified network of control from GNSS observations in Serbia. Shows the densification measurements made at "control monuments" from the CORS stations to permit local high accuracy RTK measurements within a limited area of a 10 km diameter.

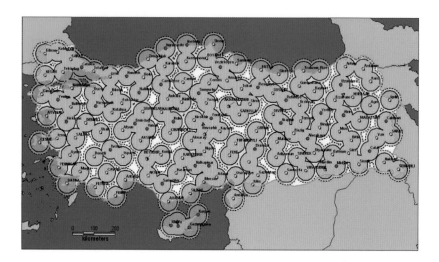

Figure 5.10   CORS network of Turkey.

GPS Satellite Geometry and GPS Software

The accuracy of GPS positioning depends on the accuracy of the range measurement, σ_R, and the satellite configurations. The positioning error, σ_P, can be expressed as

$$\sigma_P = \text{PDOP} \cdot \sigma_R$$

where

$$\text{PDOP} = \frac{1}{V}$$

with V being the volume of the tetrahedron between the satellite positions and their vectors to the receiver.

In a least squares adjustment of this geometric configuration, the covariance matrix C becomes

$$C = \sigma_R^2 (A^T A)^{-1}$$

and

$$\sigma_P^2 = \sigma_R^2 (q_{xx} + q_{yy} + q_{zz})$$

Thus, PDOP corresponds to the square root of the trace of the covariance matrix.

Each receiver manufacturer (e.g., Trimble, Leica, Ashtech, Topcon) uses its own data format for calculations of positions, including accuracy indications such as the PDOP (GP Survey for Trimble, SKI for Leica, GPPS for Ashtech).

General purpose postprocessing programs require a common translator program such as RINEX. These general purpose programs for multistation adjustments have been written by research institutes, for example, BERNESE by the University of Berne in Switzerland, GEONAP by the University of Hannover in Germany, and DIPOP by the University of New Brunswick in Canada. GEONAP uses the original phase data, and BERNESE and DIPOP use their double differences.

ACCURACY AUGMENTATION SYSTEMS

Satellite-based argumentation systems have been introduced to improve the positional accuracy of code-based GPS receivers. In North America, WAAS (Wide Area Argumentation System), and in Europe, EGNOS (European Geostationary Navigation Overlay System), have been made available since 2003 to improve the GPS positioning accuracy by code receivers to 1 to 2 m. The positioning corrections are derived from permanent GPS observation stations (25 in the United States and more than 10 in Europe) with well known portions. These observed positions generate correction signals, which are transmitted from geostationary communication satellites to the GPS code receivers, where observations are made. In Japan, MSAS has been established.

A worldwide collection system has been introduced by Fugro as Omnistar. Recently, Trimble has brought out a similar system (see Figures 5.11 and 5.12).

To improve the accuracy of GNSS, the U.S. GPS system intends to introduce a third frequency to its satellites to compensate for ionospheric signal disturbances. In addition to this, Kalman filtering is to be used for better orbit determinations of the GPS satellites. When only decimeter–ground accuracy is required, this is intended to counteract the need for the costly operation of CORS systems by the so called PPP.

GSM MOBILE PHONE LOCATION

GPS positioning depends on a direct line of sight between the satellite and the receiver; it therefore does not work in buildings or under dense vegetation. In areas where dense mobile telephone networks exist, it is possible to build up a location system that works in areas within reception reach. For this, the

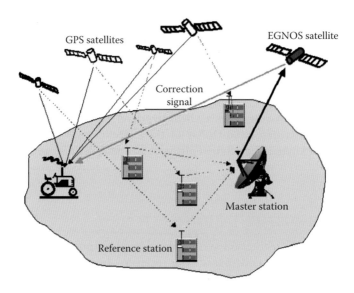

Figure 5.11   Principle of accuracy augmentation systems.

positions of the mobile phone transmission and reception antennas at which mobile phone calls are received must be known. These form the cells of the network. In urban areas, the design of these cells is on average about 700 m wide, in rural areas they can reach dimensions of tens of kilometers.

The identification of the cell is possible by the use of slightly different frequencies. Within the cell a transmitted time code permits the measurement

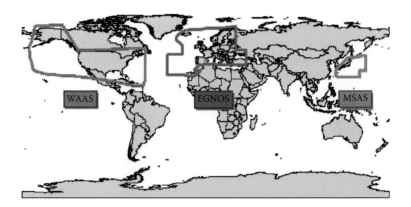

Figure 5.12   Coverage for EGNOS, MAAS, and MSAS.

of the distance between the cell antenna and the mobile phone. By intersection of at least two, three, or more distances between the cell antennas and the mobile phone, a position is determinable. Tests for uses of emergency location systems in German cities have resulted in positioning accuracies of about 100 meters.

Chapter 6

Cost Considerations

At a time when various technical possibilities are available for the acquisition, management, and updating of geoinformation, it is important to consider cost aspects. Scientists have generally been reluctant to make cost comparisons. They left suggestions to the forces of the market. But in the past, the geoinformation technologies followed slow traditional developments in the disciplines concerned. In the rapidly changing technology of today, cost considerations are important to judge which methodology is best utilized for a certain application.

There is an additional difficulty. When discussing costs, these are only rough indicators for actual prices in an open market. In general, a price for a product or for a service is composed of a cost figure plus an overhead plus the risk to get remunerated. Overheads vary greatly with salary structures of administrative services and the amount of taxes to be paid. Western Europe has considerably higher overhead percentages than Eastern Europe or countries in developing continents.

Price could be as much as 200% of cost. If the services are provided on contract to a distant country, the risk increases and prices may be as much as 300% of cost. Due to high labor costs in the developed world, automation-derived products are more favored in the service-oriented high-tech economies. For these products, lower overhead costs may be charged.

In the age of a global economy, the mapping industry is in a transient stage. It is characterized by global partnerships on the one hand, and on the other hand by an attempt to diversify the products to regional enterprises where they are produced cheapest.

A few examples of cost strategies, based on world market prices, are given here.

COSTS OF AERIAL PHOTOGRAPHY, ORTHOPHOTOGRAPHY, AND TOPOGRAPHIC LINE MAPPING

For the cost of services an assessment was made in Central Europe and in China.

Aerial photography costs consist of a base cost for mobilization of the aircraft plus a charge for the image. Mobilization of a flight will cost about $5000 in season and $3000 out of season. A cost per image of about $10 is to be added. The mobilization cost is only valid if the airport of the survey plane is reasonably close to the survey area and not if the aircraft is to be transported to a distant region.

The orthophotography production with standard aerial cameras is based on costs per image:

Production	Central Europe	China
Scanning (if required)	$15/image	$2 to $4/image
Aerial triangulation	$25/image	$8/image
Generation of a digital elevation model	$120/image	$10/image
Digital orthophoto generation	$30/image	$ 3/image
Mosaicking of digital orthophotos	$20/image	$2/image

The digital elevation model (DEM), whether automatically produced by image correlation or by measurement in stereo plotters, is the costliest part of the process. But it also generates the needed elevation base data. The price difference between Europe and China may not only reflect the labor cost differential but also the quality required.

If a digital elevation model is already available from other sources (previous topographic surveys, laser scanning, or radar interferometry), then the orthophoto process is considerably cheaper. In any case, the orthophotography production is an automated process with little cost variation around the world.

Line mapping in stereo plotters or on screen on digital workstations, however, is a labor-intensive process. Depending on the topographic detail in an image pair or on a photograph, the restitution time can vary from 10 hours per model in Europe to 8 hours in China for rural areas to 100 hours per model in urban areas in Europe to 16 hours per model in China. For line mapping, labor prices of the developing countries are much more competitive than those of developed countries, provided that the quality of the product can be maintained. Labor costs per hour can vary, for example, from $15 per hour in China or India to $40 per hour in Germany.

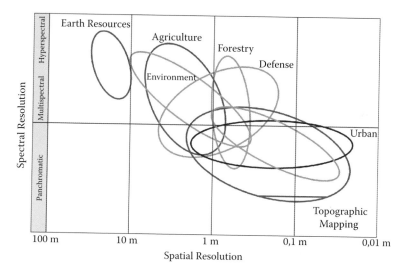

Figure 6.1   User requirements for resolution of imagery. (Drawing courtesy of LH Systems [Leica Geosystems], San Diego, California. © Leica Geosystems, 2002.)

On the basis of this analysis, the cost advantage is either in favor of outsourcing to a country with low labor costs or to push for automation of the workflows. This has been met by the development of Astrium's "Pixel Factory," or of Microsoft-Vexcel's UltraMap, which are able to automate the entire orthomapping process at high quality.

The resolution of objects, the accuracy of their geometric measurement, and the cost of the restitution are image-scale dependent. Figure 6.1 indicates the demand for panchromatic and multispectral resolution by various users.

Scale dependency versus cost is illustrated here with three examples for analogue imagery. For digital imagery the assessment may be more difficult, since the camera parameters vary greatly.

An analogue image scale of 1:6500 provides a resolution and a positional restitution accuracy of 6.5 cm. Scanned at 15 μm, it is suitable for the generation of 10 cm ground pixel orthophotos. This scale permits the mapping of utility manholes.

A photo covers an area of $a \times a = 1.5 \times 1.5$ km, and a neat model with a longitudinal overlap of 60% and a lateral overlap of 30% ($b = 0.4a$, $q = 0.7a$) covers an area of $0.28a^2 = 0.626$ km². This area can be flown at $10 plus proportional mobilization cost, and it can be produced into an orthophoto including DEM generation at $210. For an average restitution time of 50 hours per model, at $40 per hour, the line mapping cost for this area is $2000.

An urban area of 250 km² is covered by 400 photographs. The DEM has a standard deviation of ±10 cm. Therefore, the project costs become:

Aerial photography	$4000 + $4000 = $8000
Scanning	$6000
Aerial triangulation	$10,000
Digital elevation model	$48,000
Digital orthophoto generation	$12,000
Mosaicking	$8000
Total orthophoto cost	$92,000 or $368/km²
Line mapping 1:1000	$80,0000
Total line mapping cost	$892,000 or $3568/km²

If the same area is covered at an image scale of 1:13000, the resolution and the positional accuracy is 13 cm. A 20 cm ground pixel orthophoto may be generated. This scale is suitable for the mapping of buildings and road features. A photo covers an area of 3 × 3 km. The neat model area is 2503 km². The urban area of 250 km² is imaged by 100 photos. The DEM has a standard deviation of ±20 cm.

The project costs become:

Aerial photography	$4000 + $1000 = $5000
Scanning	$1500
Aerial triangulation	$2500
Digital elevation model	$12,000
Digital orthophoto generation	$3000
Mosaicking	$2000
Total orthophoto cost	$26,000 or $104/km²
Line mapping 1:2000	$200,000
Total line mapping cost	$226,000 or $904/km²

From an image scale of 1:40000, a resolution and a positional accuracy of 40 cm can be achieved. Scanned at 12.5 μm, it generates 50 cm ground pixel orthophotos. This scale satisfies planning needs. A photo covers an area of 9.2 × 9.2 km². The neat model area is 23.7 km². The urban area of 250 km² is covered by 11 photos. The DEM has a standard deviation of ±60 cm.

The project costs become:

Aerial photography	$4000 + $110 = $4110
Scanning	$165
Aerial triangulation	$275
Digital elevation model	$1320
Digital orthophoto generation	$330
Mosaicking	$220
Total orthophoto cost	$6420 or $25.68/km^2
Line mapping 1:10000	$22,000
Total line mapping cost	$28,420 or $88/km^2

The example clearly demonstrates that the orthophoto is a much cheaper line map substitute. Line maps have their advantages for topological analysis in a GIS but not all features need to be vectorized. Because of the cost element, some non-European countries prefer to restrict their vector database to the property cadastre. All relevant attributes can be attached to the parcel. Due to the importance and the sensitivity of land transactions, the cadastre including the attributes can be maintained on a real-time basis for each transaction. The cadastral vector database may be overlaid with the orthophoto in raster form to depict buildings and road features. If desired, buildings may be numbered and identified by pointers to attach attributes, without the need to show the outlines in vector graphics.

The utility network is essentially a line network between manholes, which can easily and rapidly be measured by Real-Time Kinematic (RTK) GPS surveys as an additional vector layer for each utility type. Like the property cadastre, it can also be superimposed with the orthophoto. In surveying the utility manholes, their attribute information may be verified or added on-site. As the number of utility points for each utility is rather small, ground surveys are more effective and less costly.

GPS-SUPPORTED GROUND SURVEYS

Ground surveys are generally carried out for cadastral purposes of larger areas in terrain, where visibility requirements of the boundaries prohibit the use of photogrammetric techniques. They may also be preferable in countries where photo-adjudication is not accepted due to accuracy concerns.

In Europe, cadastral data are already existent. Therefore, a survey cost comparison with photogrammetric methods is not useful there. However, in a number of development projects, ground survey costs per parcel, including the land registration or land titling aspects, have been established.

In Albania, a new cadastre has been established by European funding at a cost of $5 per parcel. The procedure used was aerial photography–aerial triangulation–digital elevation model–digital orthophoto generation followed by a public photo adjudication process.

In Georgia, a large German technical cooperation project was carried out by GPS-supported electronic tacheometers at a cost of $10 per parcel. A survey crew is able to measure about 50 parcels per day. In doing so, it has been proved useful to support the ground surveys with aerial photos or orthophotos. For this purpose, "digital plane tables" in the form of large-screen PDAs may be used to record the measured GPS or electronic tacheometer measurements on the screen. These data are superimposed with preprepared (ortho-)photographic data on the screen, which helps to identify points to be measured terrestrially. A digital plane table costs about $10,000.

An urban area of 250 km² has about 80,000 parcels. The cost of surveying these terrestrially would therefore be $800,000, which is about the same as the photogrammetric line mapping cost at the scale of 1:1000, but about 4 times as much as the photogrammetric mapping cost at the scale of 1:2000. This corresponds to a terrestrial survey cost of $3200/km².

DIGITAL ELEVATION MODELS

For the acquisition of digital elevation models, there exist a great variety of technical possibilities, as shown in Table 6.1.

AERIAL TRIANGULATION VERSUS DIRECT SENSOR ORIENTATION

Another issue is the possibility of replacing the photogrammetric aerial triangulation process with the possibility of measuring sensor positions and sensors directly by GPS/IMU (inertial measuring unit) systems. The combined use of aerial triangulation and GPS/IMU observations in a bundle block adjustment is, in any case, advantageous for the limitation of the needed ground control points and for the increased quality control possibilities for improved reliability of geocoding.

But the acquisition of additional inertial sensor data significantly increases the investment costs for aerial photography. The same is true in the case where inertial sensor-dependent digital airborne scanners are intended to be used instead of standard aerial photography.

TABLE 6.1   COMPARISON OF ACQUISITION COSTS FOR DIGITAL ELEVATION MODELS

Methodology	Height Accuracy	Cost	Application
Satellite radar interferometry	±12 m	$2/km²	Topography
Airborne radar	±2 m	$20/km²	Topography
Aerial photography 1:40000	±0.20 m	$25/km²	Topography, city models
Aerial photography 1:13000	±0.20 m	$100/km²	Nonforested areas
Aerial photography 1:6500	±0.10 m	$350/km²	Nonforested areas
Airborne laser	±0.15 m	$500/km²	Forest, building heights, coastal areas
Ground surveys	±0.10 m	$1000/km²	Dense forest

An airborne inertial sensor system will require the acquisition of a $500,000 investment. An inertial measurement unit by Applanix or IGI will cost between $500,000 and $250,000 as an investment. It needs to be balanced against the savings obtained by skipping aerial triangulation. This depends on the volume of work anticipated to justify these investment costs.

MAPPING FROM SPACE

Ever since Landsat TM initiated the provision of satellite imagery at 30 m ground pixels in 1982, there have been attempts to commercialize the market for space imagery. The U.S. Landsat Commercialization Act of 1984 paved the way for the marketing of space images, particularly for 10 m resolution panchromatic Spot images by Spot-Image. Whereas Landsat images covering 185 km × 185 km areas at 30 m pixels and Spot images covering 60 km × 60 km areas at 10 m in black-and-white and 20 m in color would cost several thousand dollars each, the prices for image products have been considerably lowered since. The recent lowering of Landsat prices to a few hundred dollars per scene, following the public domain principles of the U.S. government, has brought a disturbance in the imagery market of medium-resolution satellites.

TABLE 6.2   CARTERRA-IKONOS 2 PRODUCTS BY SPACE IMAGING

Type of Data	Geometric Accuracy	Price (November 2001)
Raw data	±12 m	$29/km^2
Orbit corrected data	±6 m	$50/km^2
Geocorrected data on the basis of ground control	±3 m	$100/km^2
Stereo imagery	±1 m	$200/km^2

With the advent of high-resolution satellites by Ikonos 2 in 1999, with 1 m pixels in black-and-white and 4 m in color, the Space Imaging company now provides a number of products as listed in Table 6.2.

The Ikonos 2 products now compete with Russian space photography products of the Spin 2 program (KVR 1000 panoramic photographs digitized to 2 m pixels) with a geometric accuracy of ±5 m at $25/km^2.

Another advantageous alternative is the use of standard aerial photography at the scale of 1:40000 with a much higher resolution of 0.5 m pixels at about $25/km^2.

Urban centers with a rapidly growing population are usually faced with inadequate and not up-to-date mapping coverage, such as in the city of Tirana, Albania, which grew from a population of 200,000 in 1980 to 600,000 in 2005. The result was illegal construction of residential buildings at a time when it became impossible to renew the cadastral and planning information commensurate with the rapid changes. In less than 10 years an area of less than 5% of the area of 60 km^2 was completed for an affordable sum of US$10 million.

A project in 2006 supported by the World Bank, however, permitted to supply reasonable planning information for the 60 km^2 area for a relatively small sum of US$200,000 by acquiring one Quickbird high-resolution satellite image and an ArcGIS, and by creating a GIS database for the urban planning purposes of Tirana.

In the geocoded satellite image all new buildings were digitized as vectors (Figure 6.2). The buildings were displayed in red (Figure 6.3). The entrances to each building were geocoded using GPS code observations with an ArcPad with corrections from the accuracy augmentation system EGNOS. At each house entrance the required attribute information about population and facilities was collected (see Figure 6.4 for the display of house entrances).

To include utility information, the utility plans from different service providers were scanned and geocoded (Figure 6.5).

Figure 6.2   Digitization of building outlines from Quickbird image in Tirana.

Figure 6.3   Display of buildings in ArcGIS in red.

Figure 6.4   GPS augmented accuracy surveys of house entrances, where attributes were collected.

Figure 6.5   Scanned utility maps.

Figure 6.6   Display of land use.

Other information, such as land use (Figure 6.6), the existing parking facilities (Figure 6.7), the installed traffic lights (Figure 6.8), and the boundaries of urban administration districts (Figure 6.9) could be superimposed on the geocoded satellite image. On the basis of subsequent new satellite images a year later the updating of new buildings erected in the meantime became possible (Figure 6.10).

Figure 6.7   Parking facilities.

Figure 6.8   Traffic lights.

AUTOMATED FEATURE EXTRACTION

Automated DTM collection has been realized as an operational measure since 1990. At the same time, attempts began for automatic feature extraction sponsored by DARPA, the Defense Advanced Projects Agency of the United States, with Intergraph winning the contract.

While eCognition software by Definiens became more successful to classify objects from multispectral images by segmentation rather than by traditional

Figure 6.9   Urban administration districts.

Figure 6.10   Updating of buildings by new imagery superimposed with old building database.

multispectral classification approaches on a pixel basis, the concentration was on feature extraction of objects, such as roads, buildings, drainage patterns, and land cover.

The object's size, shape, texture, pattern, its shadow, and the spatial context needed to be formulated with tools developed by the computer vision community, for example, the use of snakes and dynamic programming for the detection and location of road networks.

These tools were highly successful for change detection, but they could not improve the skills of a human analyst operator beyond a success rate of 80% thus far.

Nevertheless, this gave rise to an increase in the efficiency of feature extraction by human operators by a factor of 3 through automation, when a quality assessment of the feature extraction was included. This helped to identify areas which needed to be checked by human analysts.

This procedure has become especially useful in verifying the currency of geospatial databases from more recent imagery. In this way, areas in need of updating can be identified by automatic procedures.

Chapter 7

Technological Changes

A more philosophical approach to technological changes has been initiated by the studies of Oswald Spengler of Munich (1880–1936). His treatise *Cultural Cycles of Civilizations*, in which he described the cycles in the development of civilizations—blossoming, maturity, and decay—inspired the British historian Arnold Joseph Toynbee (1889–1975) to write *A Study of History* in which he described the cycles of civilizations from their early beginnings to the present. The motivations for changes were *challenge* and *response.* Toynbee's ideas affected the studies of Pitirim Alexandrovich Sorokin (1889–1968), a Harvard professor, who wrote *Social and Cultural Dynamics.*

A technological link to these cycles was established by Nikolai Dmitriyevich Kondratjev (1892–1930), the founder of the Conjuncture Economics Institute in Moscow, the originator of the 5-year plan for Soviet agriculture. His treatise *The Major Economic Cycles* (1926) postulates that economic changes are generated by new technology. In Figure 7.1, these economic cycles at about 50-year periods are generated by technological inventions. This initiates a rapid growth of the technology and its application.

It is of interest that, for example, the four stages of the development of the geoinformation discipline photogrammetry also fall into this pattern:

- Single image photogrammetry, from 1850 to 1900
- Analogue stereo photogrammetry, from about 1900 to 1950
- Analytical photogrammetry, from about 1950 to 2000
- Digital photogrammetry, for the 21st century

The geoinformation disciplines of the 20th century have favored an ever-increasing specialization. In the 21st century, a renaissance of thought is in order, in which the various specializations must be considered together and used jointly for a sustainable development.

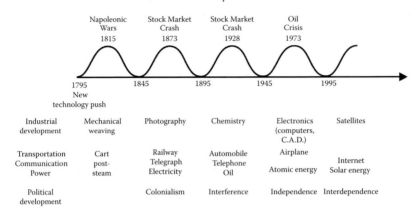

Figure 7.1   Kondratjev's economic cycles.

Bibliography

CHAPTER 1—INTRODUCTION

Capters, F. and Decleir, H. *The World in Perspective: A Directory of World Map Projections*, John Wiley, New York, 1990.

Everest, G. *An Account of the Measurement of an Arc of the Meridian between the Parallels 18°3′ and 24°7′*, London, 1830.

Helmert, F.R. *Die mathematischen und physikalischen Theorien der höheren Geodäsie*, 1880, reprint Minerva, Frankfurt, 1961.

Konecny, G. "Paradigm Changes in ISPRS from the First to the Eighteenth Congress in Vienna," *ISPRS Highlights*, 1 (1996), 10–15.

Konecny, G. "Photogrammetry and Remote Sensing in Transition to Geoinformatics," *PGF* 6 (1998), 329–335.

Konecny, G. "Status of Mapping in the World," ISPRS publication to UN-GGIM3 Conference, Cambridge, England, 2013.

Longley, P.A., Goodchild, M.F., Maguire, D.J., and Rhind, D.W. *Geographic Information Systems*, John Wiley, New York, 1999.

Seeber, G. *Satellite Geodesy*, De Gruyter, Berlin, 1993.

Smith, J.R. *Introduction to Geodesy: The History and Concepts of Modern Geodesy* (Series in Surveying and Boundary Control), Wiley, New York, 1997.

Torge, W. *Geodesy*, 3rd edition, De Gruyter, Berlin, 2001.

UN Secretariat. "The Status of World Mapping 1990," *World Cartography*, 1993, UN Cartographic Conference, Beijing, paper prepared by A. Brandenberger and S. Ghosh.

United Nations Statistic Division. *World Statistics Pocketbook*, Series IV, No. 21, UN Publications, New York.

CHAPTER 2—REMOTE SENSING

American Society of Photogrammetry and Remote Sensing. *Manual of Remote Sensing*, American Society of Photogrammetry, Falls Church, VA, 1975.

Askne, J. (ed.). *Sensors and Environmental Applications of Remote Sensing*, Proceedings of the 14th EARSeL Symposium, Göteburg, Sweden, Ashgate Publishing Company and A.A. Balkema International Publishers, Rotterdam and Brookfield, VT, 1995.

Campbell, J.B. *Introduction to Remote Sensing*, Taylor & Francis, London, 1996.

Canada Centre for Remote Sensing. *Fundamentals of Remote Sensing*, tutorial, http://www.cors.nrcan.gc.ca/ccrs/eduref/tutorial/tutore.html.

Canty, M.J. *Fernerkundung mit neuronalen Netzen*, Expert Verlag Remingen-Matensheim, Germany, 1999.

CNES. "Resources in Earth Observation," *Case Studies, Data and Information for Education and Developing Countries*, 1st edition, CSIRO and Smith System Engineering Ltd, http://sirius-ci.cst.cnes.fr:8100/CD-ROM-97/astart.letm.

Colwell, R.N., et al. "Basic Matter and Energy Relationships Involved in Remote Reconnaissance," *Photogrammetric Engineering*, 29 (1963), 761–799.

Congalton, R.G. "A Review of Assessing the Accuracy Classification of Remotely Sensed Data," *Remote Sens. Environ.*, 27 (1991): 35–46.

Cracknell, J.P., and Hayes, L.W.B. *Introduction to Remote Sensing*, Taylor & Francis, London, 1991.

Dowman, I., Jacobsen, K., Konecny, G., and Sandau, R. *High Resolution Optical Satellite Imagery*, Whittles Publishing, Caithness, U.K., 2012.

Gibson, P., and Power, C. *Introductory Remote Sensing: Principles and Concepts*, Taylor & Francis, London, 2000.

Hapke, B. *Theory of Reflectance and Emittance Spectroscopy*, Cambridge University Press, Cambridge, 1993.

Henderson, F.M., and Lewis, A.J. "Principles and Applications of Imaging Radar," *Manual of Remote Sensing*, 3rd edition, vol. 2, John Wiley, New York, 1998.

Jacobsen, K. "High Resolution Satellite Imaging Systems—Overview," *PGF* 6 (2005), 487–496.

Jacobsen, K. "Characteristics of Worldwide and Nearly Worldwide Height Models," *ISPRS WG IV/2 Workshop Interexpo Geo-Siberia, Novosibirsk* 2013, proceedings, 42–57.

Jacobsen, K. "Characteristics of Very High Resolution Satellites for Topographic Mapping," *ISPRS WG IV/2 Workshop Interexpo Geo-Siberia, Novosibirsk* 2013, proceedings, 58–78.

Lillesand, T.M., and Kiefer, R.W. *Remote Sensing and Image Interpretation*, John Wiley, New York, 1979.

Lippmann, R.P. "An Introduction to Computing with Neural Nets," *IEEE ASSP Magazine*, April 1987, 4–22.

Milman, A.S. *Mathematical Principles of Remote Sensing*, Sleeping Bear Press, Chelsea, MI, 1999.

Nagler, T., and Rott, H. *Overview of Current and Planned Spaceborne Earth Observation Systems*, Space Application Institute, Joint Research Centre, Ispra, 1998, EUR 18673 EN.

Pelzer, H., Crampéa, F., and Rosen, P. "The Mir 7.1, Hector Mine, California Earthquake: Surface Rupture, Surface Displacement Field and Fault Ship Solution from ERS SAR Data," *Earth and Planetary Sciences*, 333 (2001), 545–555.

Rencz, A.N., and Ryerson, R.A. (eds.). *Manual of Remote Sensing, Vol. 3: Remote Sensing for the Earth Sciences*, 3rd edition, John Wiley, New York, 1999.

Richards, J.A., and Xinping, J. *Remote Sensing Digital Image Analysis*, 3rd edition, Springer, Berlin, 1998.

Sabins, F.F. *Remote Sensing: Principles and Interpretation*, 3rd edition, W.H. Freeman, New York, 1996.

Schanda, E. *Physical Fundamentals of Remote Sensing*, Springer Verlag, Berlin and Heidelberg, 1986.

Schowenderdt, R.A. *Remote Sensing: Models and Methods for Image Processing*, 2nd edition, Academic Press, San Diego, CA, 1997.

Short, N.M. Sr. *The Remote Sensing Tutorial: An Online Handbook at NASA's Goddard Space Flight Center*, http://rst.gsfc.nasa.gov.

SPIE. *Proceedings of the Conference on Commercial Remote Sensing Platforms and Applications*, International Society for Optical Engineering, Bellingham, WA, 1999.

Spiteri, A. (ed.). *Remote Sensing: Integrated Applications for Risk Assessment and Disaster Prevention for the Mediterranean*, Proceedings of the 16th EARSeL Symposium, Ashgate Publishing Co. and A.A. Balkema, Rotterdam, Brookfield, VT, 1997.

Steinborn, W., and Sprengelmeier-Schnock, I. (eds.). *Raumfahrt zum Nutzen Europas: Die Perspektiven der Fernerkundung mit Satelliten*, Herbert Wichmann Verlag, Karlsruhe & Heidelberg, 1993.

CHAPTER 3—PHOTOGRAMMETRY

Ackermann, F. "Airborne Laser Scanning: Present Status and Future Expectations," *ISPRS Journal of Photogrammetry and Remote Sensing*, 54 (1999), 2–3.

Ackermann, F., Ebner, H., and Klein, H. "Ein Rechenprogramm für die Streifentriangulation mit unabhängigen Modellen," *Bildmessung und Luftbildwesen*, 1970, 206–217.

Alobeid, A., Jacobsen, K., and Heipke, C. "Comparison of Matching Algorithms for DSM Generation in Urban Areas from Ikonos Imagery," *Photogrammetric Engineering and Remote Sensing*, 76 (2010), 1041–1050.

Baltsavias, E.P. "Airborne Laser Scanning: Basic Relations and Formulas," *ISPRS Journal of Photogrammetry and Remote Sensing*, 54 (1999), 199–214.

Ebner, H. "Zwei neue Interpolationsverfahren und Beispiele für ihre Anwendung," *Bildmessung und Luftbildwesen*, 1979, 15–27.

Förstner, W. "Probabilistic Data Analysis Using Graphical Models," Lectures at Leibniz University Hannover (PowerPoint), April/May 2013.

Fraser, C.S., Dial, G., and Grodecki, J. "Sensor Orientation via RPCs," *ISPRS Journal of Photogrammetry and Remote Sensing*, 60 (2006), 182–194.

Fricker, P. "ADS 40: Progress in Digital Aerial Data Collection," *Photogrammetric Week*, 2001, 105–116.

Greve, C. (ed.). "Digital Photogrammetry," an addendum to the *Manual of Photogrammetry*, American Society of Photogrammetry and Remote Sensing, Bethesda, MD, 1996.

Grodecki, J., and Dial, G. "Block Adjustment of High Resolution Satellite Images Described by Rational Polynomials," *Photogrammetric Engineering & Remote Sensing*, 69 (2003), 59–68.

Heipke, C., 2001, "Digital Photogrammetric Work Stations: A Review of the State of the Art for Topographic Applications," *GIM International* 15 (2001), 35–37.

Heipke, C., Jacobsen, K., and Wegmann, H. *The OEEPE-Test on Integrated Sensor Orientation: Analysis and Results*, OEEPE Workshop Integrated Sensor Orientation, Hannover, September 2001, 2002 OEEPE Official Publication 43, 31–49.

Helmholz, P., Becker, C., Breitkopf, U., Büschenfeld, T., Busch, A., Braun, C., Grünreich, D., et al., "Semi-Automatic Quality Control of Topographic Data Sets," *Photogrammetric Engineering and Remote Sensing* 78 (2012), 959–972.

Hinz, A., Dörstel, C., and Heier, H. "DMC: The Digital Sensor Technology of Z/I Imaging," *Photogrammetric Week*, 2001, 93–104.

Hirschmüller, H. "Accurate and Efficient Stereo Processing by Semi-Global Matching and Mutual Information," *Proceedings of IEEE Conference on Computer Vision and Pattern Recognition*, 2, 807–814.

Hobrough, G.L. "Automatic Stereo Plotting," *Photogrammetric Engineering*, 1959, 763–769.

Jacobsen, K. "Sensororientierung" in Sandau, R. (ed.), *Digitale Luftbildkamera*, Herbert Wichmann Verlag, 2005, 118–126.

Jacobsen, K., Cramer, M., Ladstätter, R., Ressl, C., and Spreckels, V. "DGPF Project: Evaluation of Digital Photogrammetric Camera Systems: Geometric Performance," *PFG* 2 (2010), 85–98.

Konecny, G., and Lehmann, G. *Photogrammetry*, De Gruyter, Berlin, 1984.

Konecny, G., Kruck, E., and Lohmann, P. "Ein universeller Ansatz für die Geometrische Auswertung von CCD-Zeilenabtasteraufnahmen," *Bildmessung und Luftbildwesen*, 54 (1986), 139–146.

Konecny, G. *The International Society for Photogrammetry and Remote Sensing: 100 Years of the Society*, ISPRS, BEV Vienna.

Kraus, K. "Interpolation nach kleinsten Quadraten in der Photogrammetrie," *Bildmessung und Luftbildwesen, Journal of the German Society of Photogrammetry*, 1972, 7–12.

Kraus, K. *Photogrammetry*, 4th edition, Dümmler Verlag, Stamm GmbH, Köln, 1994.

Kremer, J. "CCNS and Aerocontrol: Products for Efficient Photogrammetric Data Collection," *Photogrammetric Week*, 2001, 85–92.

Luhmann, T. *Nahbereichsphotogrammetrie*, Herbert Wichmann, Heidelberg, 2000.

McKeown, D.M. Jr. "Toward Automatic Cartographic Feature Extraction," *Mapping & Spatial Modelling for Navigation*, NATO ASI Series, 65 (1990), 149–180.

Mikhail, E., Bethel, J.S., and McGlone, C. *Introduction to Modern Photogrammetry*, John Wiley, New York, 2001.

Mostafa, M., Hutton, J., and Raid, B. "GPS/IMU Products: The Applanix Approach," *Photogrammetric Week*, 2001, 63–84.

Ok, A.O., Wegner, J.D., Heipke, C., Rottensteiner, F., Soergel, U., and Toprak, V. "Matching of Straight Line Segments from Aerial Stereo Images of Urban Areas," *ISPRS Journal of Photogrammetry and Remote Sensing*, 74 (2012), 133–152.

Petrie, G. "3D Stereo Viewing of Digital Imagery," *Geoinformatics*, 4 (2001), 24–29.

Read, R., and Graham, R. *Manual of Aerial Survey: Primary Data Acquisition*, Whittles Publishing, Caithness, 2001.

Schenk, T. *Digital Photogrammetry*, vol. 1, Terra Science, Laurelville, OH, 1999.

Schroeder, M. *Zur photographischen Aufnahmetechnik im Weltraum für kartographische Aufgaben*, Wiss. Arbeiten der Fachrichtung Vermessungswesen, Universität Hannover, Nr. 165.

Scott, C. *Introduction to Optics and Optical Imaging*, John Wiley, New York, 1997.

Seitz, S. "Landmarks in 3D Computer Vision," NSF Workshop on Frontiers of Computer Vision, August 21, 2011, University of Washington.

Tao, C.V., Hu, Y., and Jiang, W. "3D Reconstruction Methods Based on Rational Function Model," *Photogrammetric Engineering and Remote Sensing*, 68 (2002), 705–714.

Slama, C.C. (ed.). *Manual of Photogrammetry*, 4th edition, American Society of Photogrammetry and Remote Sensing, Falls Church, VA, 1980.

CHAPTER 4—GEOGRAPHIC INFORMATION SYSTEMS

AMD. *The AMD x86-64 TM Architecture*, Programmers overview, Publication 24108, January 2001.

ATKIS. *Amtliches Topographisch-kartographisches Informationssystsem*, http:// www.atkis.de.

Bartelme, N. *Geoinformatik*, 3rd edition, Springer, Berlin–Heidelberg–New York, 2000.

Berry, J.K. *Beyond Mapping: Concepts, Algorithms and Issues in GIS*, John Wiley, New York, 1996.

Bill, R. *Grundlagen der Geoinformationssysteme*, vol. 2, Wichmann, Heidelberg, 1991.

Bill, R., and Fritsch, D. *Grundlagen der Geo-Informationssysteme*, vol. 1, Wichmann, Karlsruhe, 1991.

Chou, Y.-H. *Exploring Spatial Analysis in GIS*, OnWord Press and Delmar, Thomson Learning, Albany, NY, 1996.

de By, R.A., et al. *Principles of Geographic Information Systems: An Introductory Textbook*, International Institute of Aerospace Survey and Earth Sciences (ITC), Enschede, 2001.

de Mers, M.N. *Fundamentals of Geographic Information Systems*, John Wiley, New York, 1999.

ESRI. *Map Objects Internet Map Server Reference* (issued with software), ESRI, Redlands, CA, 1998.

FIG. *Reforming and Benchmarking the Cadastre: Measuring the Success*, Proceedings of the FIG Comm 7 Symposium in Gävle, June 2001.

Gerke, M., and Heipke, C. "Image-Based Quality Assessment of Road Data Bases," *International Journal of Geoinformation Science*, 8 (2008), 871–894.

Gifford, F. "Internet GIS Architecture: Which Side is Right for You?" *GeoWorld*, 1999.

Heipke, C. *GIS Course Rawalpindi 1999* (Components of GIS Hardware, Components of GIS Software, Introduction to GIS Data Bases), CD-ROM Institute for Photogrammetry and GeoInformation, University of Hannover.

Kaufmann, J., and Steudler, D. *Cadastre 2014: A Vision for a Future Cadastral System*, FIG 1998.

Konecny, G. "Photogrammetry and Remote Sensing in Support of GDI," in Groot, R., and McLaughlin, J. (eds.), *Geospatial Data Infrastructure*, Oxford University Press, 2000, 195–216.

Konecny, G. "Cadastral Mapping with Earth Observation Technology," in Li, D., Shan, J., and Gong, J., *Geospatial Technology for Earth Observations*, Springer, New York, chapter 15.

Korn, G.A., and Korn, T.M. *Mathematical Handbook for Scientists and Engineers*, McGraw Hill, New York–Toronto–London, 1961.

Korte, G. *The GIS Book*, 5th edition, OnWord Press and Delmar, Thomson Learning, Albany, NY, 2000.

Linder, W. *Geo-Informationssysteme: Ein Studien- und Arbeitsbuch* (in German), Springer, Berlin, 1999.

Longley, P., and Batty, M. *Spatial Analysis: Modelling in a GIS Environment*, John Wiley, New York, 1997.

Marshall, J. "Developing Internet-Based GIS Applications," http://www.giscafe.com/ TechPapers/Papers/paper 058/p405.htm.

Mitchell, A. *The ESRI Guide to GIS Analysis, Vol. 1: Geographic Patterns and Relationships*, ESRI Press, Redlands, CA, 1999.

Murai, S. *Textbook on Remote Sensing and GIS*, 1999. "Remote Sensing Notes" produced by the National Space Development Agency of Japan (NASDA). "GIS Workbook" produced by the Asian Center for Research and Remote Sensing (ACRoRS) at Asian Institute of Technology (AIT).

National Spatial Data Infrastructure. "The 2002 National Spatial Data Infrastructure Cooperative Agreements Programme (CAP)," http://www.fgdc.gov.

Standish, T.A. *Data Structure Techniques*, Addison-Wesley, Reading, CA, 1980.

Sylla, C.I., Wade, I.A., Hengue, P., and Gerbe, E. "Environmental Information Systems in Sub-Saharan Africa," Country Case: Senegal, Report published by the GTL, Eschborn, Germany, 1997.

Winget Godin, L. *GIS in Telecommunications Management*, ESRI Press, Redlands, CA, 2000.

Zeller, M. *Modelling Our World*, ESRI Press, Redlands, CA, 1999.

CHAPTER 5—POSITIONING SYSTEMS

Kaula, W.M. *Theory of Satellite Geodesy*, Dover Publications, Mineda, NY, 2000.

Seeber, G. *Satellite Geodesy: Foundations, Methods and Applications*, De Gruyter, Berlin, 2000.

Teunissen, P.J.G., and Kleusberg, A. (eds.). *GPS for Geodesy*, 2nd edition, Springer, Berlin–Heidelberg–New York, 1998.

CHAPTER 6—COST CONSIDERATIONS

Betz, R., and Schott, B. *Mehr Umsatz mit Perfect-Marketing*, Max Schirm Verlag, Germany, 1996.

Gruen A., and Haihong, Li. "Semi-Automatic Linear Feature Extraction by Dynamic Programming and LSB-Snakes," *Phot. Engineering and Remote Sensing*, 63 (1997), 985–995.

Heipke, C., Woodsford, P.A., and Gerke, M. "Updating Geospatial Databases from Images," in Li, Z., Chen, J., Baltsavias, E. (eds.), *Advances in Photogrammetry, Remote Sensing and Spatial Information Sciences*, Taylor & Francis, London, 2008, 355–362.

Kern P. "Automated Feature Extraction: A History," MAPPS/ASPRS 2006 Fall Conference, San Antonio, TX.

Quakenbush, L.J. "A Review of Techniques for Extracting Linear Features from Imagery," *Phot. Engineering and Remote Sensing*, 70 (2004), 1383–1392.

Ravanbakhsh, M., Heipke, C., and Pakzad, K. "Road Junction Extraction from High Resolution Aerial Imagery," *Photogrammetric Record*, 23 (2008), 405–423.

Index